Aviation Engineering

Navigating Through the Golden Years!

Aviation Engineering

Navigating Through the Golden Years!

BY

MARIJAN JOZIC

Warrendale, Pennsylvania, USA

400 Commonwealth Drive
Warrendale, PA 15096-0001 USA
E-mail: CustomerService@sae.org
Phone: 877-606-7323 (inside USA and Canada)
             724-776-4970 (outside USA)
FAX: 724-776-0790

Library of Congress Catalog Number 2023931591
http://dx.doi.org/10.4271/9781468605396

ISBN-Print   978-1-4686-0538-9
ISBN-PDF   978-1-4686-0539-6
ISBN-ePub   978-1-4686-0540-2

To purchase bulk quantities, please contact: SAE Customer Service

E-mail: CustomerService@sae.org
Phone: 877-606-7323 (inside USA and Canada)
             724-776-4970 (outside USA)
Fax: 724-776-0790

Visit the SAE International Bookstore at books.sae.org

Publisher
Sherry Dickinson Nigam

Development Editor
Amanda Zeidan

Director of Content Management
Kelli Zilko

Production and Manufacturing Associate
Brandon Joy

"Who else could be most qualified to write a book about the golden years of aviation than Marijan Jozic?

From the bowels of electromechanical instrumentation to the latest flight management computers, from the 'steam gauges' to LCD and Head Up displays, Marijan has seen, designed and managed their implementation. Thus then, who best to lead you in a journey through those golden years."

Randolph Johnstone
PhD, Former Boeing Associate Technical Fellow

"From the first moment I met Marijan at an ARINC Industry Activities meeting, I knew this was someone I could learn from based on his expansive knowledge and experience in aviation. Knowing how challenges were successfully overcome in the past provides encouragement and a framework for solving the issues of today and the future. I am so glad he is sharing that wealth of history, wisdom, and wit with this new book!"

Laurie Strom
Former Executive Vice President & Chief Operating Officer, SAE-ITC

"I have cherished our friendship of 20+ years and admired Marijan's leadership skills as Chairman of the Airline Maintenance Conference. For MJ truly believes in Make Avionics Great Again."

Robert Weber
Owner of R C Weber Associates,
Aviation & Aerospace Consultant
FAA DER

"A genius with passion who masterly turns technical jargon into fascinating stories."

Paul Sun

CEO of AFI KLM E&M Components Shanghai China

"Marijan is the most influential avionics expert I have ever met. His love, leadership and dedication to the industry led to many innovations."

John Leslie

Chairman, NAASCO

"His knowledge, networks, and writings truly inspire me as an Engineer and to become a Technopreneur. Marijan is my best mentor."

Tito Luthfan Ramadhan

Software Engineer and Innovation Team Leader at GMF AeroAsia

"In 2004, I was working as a consultant for CMC Electronics, a Canadian based avionics company. During the AEEC annual meeting, which was in Paris that year, one of CMC's marketing folks, John Barker and I played rock music at the CMC hospitality suite. It was such a huge success that the organization asked us to do it again the following year in Seattle, WA. That was when we met Marijan. While he was working for KLM airlines at the time, he was most enthusiastic about our rock event and asked if he could sing a few songs. We gladly accepted and the event was once again a hit. His favorite was "Proud Mary" by Creedence Clearwater Revival, and he quickly found another singer, who worked for Continental Airlines.

This group came to be known as the CMC Music Company and played at a number of these events in the years that followed. What a hoot – great to have you as our lead singer Marijan."

Phil Moylan
President at Moylan Marketing Consultants, Inc

John Barker
Sales Director

"I am one of the KLM cabin attendants who temporary worked at Marijan's department in the Avionics shop. Together with few other cabin attendants we helped engineers during the TTE project. I believe that we were following ARINC 668 specification. Just in a few days we became part of the great team with a lot of nice and helpful engineers. Finally, we saw where they repair the coffee makers, ovens, crew rest dimmers and many other familiar things. It was nice to experience how life is in the technical department. Marijan used to call us "Stess" and we got our name back at the farewell reception."

Annet Leenders
KLM Cabin Attendant

Contents

Acknowledgments

The book in your hand is worth nothing if you don't read it. It's just a dead letter on a dead tree. As an avionics engineer, I feel that I am a citizen of the world, because a good engineer operates globally. I don't want to publish this book without mentioning that I enjoyed working with my fellow engineers from around the world at: Boeing (Long Beach, Everett, Renton, Issaquah, and Wichita), CMC Electronics (Montreal and Ottawa), Honeywell (Phoenix), Rockwell Collins (Cedar Rapids and Melbourne, Florida), Smiths (Cheltenham), ECS (Franklin), EMTEQ (Miramar FL), Litton (London, Los Angeles) Canard (Minneapolis), Hollingshead (Los Angeles), Spherea (Tolouse, France), ARINC (Annapolis), and so on. Many engineers working at other airlines also inspired me. These airlines include: Atlas Air, British Airways, Northwest Airlines, Lufthansa, Focus Air Cargo, Austrian Airlines, Swissair, Nationwide Airlines, Cathay, Dragon, Japan Airlines, All Nippon Airways, MK Airlines, UPC, Cargolux, Air France, LAN, United Airlines, Delta, Garuda, Royal Air Maroc, American Airlines, FedEx, Croatia Airlines, Alitalia, Transaero, Martin Air, Transavia, Saudia, and Jat Airways, among others. Of course, some skilled KLM engineers and very knowledgeable technicians also inspired me to write some of these chapters.

I was often inspired by the strange, sometimes stupid behavior of people in and outside my own company. It is sometimes funny and rewarding (and at times frustrating) to observe the effects of poor decisions on aviation. Even such difficult situations can be inspiring, because these are often the moments from which I learned the most and which also provided motivation for writing this book.

I would like to mention one person who helped me overcome my fears and start writing: Roger Goldberg. He suggested at the AMC conference in Hamburg back in the year 2000, that it might be a good idea for me to write an article for *Plane Talk*, which he edited at the time. Three months later, I sent him my first article, "Is It Fun to Be an Engineer?" He was excited to place it in the magazine. Two days after publication, I received an e-mail from an editor of *Aviation Maintenance*. He asked me if I would like to write for him as well. Suddenly, I had two magazines to write for! For the articles published in those two magazines, I was nominated for Aerospace Journalist of the Year. My greatest thanks go to Roger, who made me a storyteller.

My family (the Dutch part) was patient with me and provided the support that enabled me to finish the book. Thank you all.

I will always have the greatest respect for all the good people working in aviation, making the world fleet safe and keeping the aircraft flightworthy. I am proud that I was part of it for more than 40 years.

1

Torchbearer

The book in your hands is worth nothing if you don't read it. The people around me are partly responsible for the book, because they encouraged me in one way or another to put my thoughts on paper. Many engineers contributed to this book, and they don't even know it. I have great respect for all of the good men and women working in aviation, making the world fleet safe and keeping the aircraft in good shape. It feels great to be part of that team.

It is often wise to start with *why*. *Why* did I decide to write? I always challenge the status quo. I want to do things differently—and preferably better—than others. I had a great career in aviation working for an airline. I could have just gone with the flow and accomplished very little. But I decided to challenge myself. Now I am challenging myself again by writing a book about the last 40 years. Forty years is a solid period of time. I am convinced that the years between 1980 and 2020 were the most exciting years in aviation history—and not just because my career spans that period. Before 1980, people were building aircraft and learning about flight. After 1980, we were working on perfecting aircraft machinery and, above all, making it safe. We started with McDonnell Douglas DC-9s, Airbus A300s, and Boeing 737-200s, and now, after 40 years, we are building very sophisticated aircraft like Boeing 787s and Airbus A350s. The modern aircraft are actually flying computers in the fuselage of composite material. Before 1980, there were many fatal plane crashes. But in these 40 years, we managed to reduce the number of accidents to nearly zero. We even had a few years without any accidents.

In 1980, I took the torch from engineers who were finishing their careers and retiring. For the next 40 years, I carried the torch. I *proudly* carried the torch. And I was the torchbearer for 40 years. Many other dedicated engineers did the same (Figure 1.1).

FIGURE 1.1   I took the torch from unknown engineers and carried it for the next 40 years.

lazyllama/Shutterstock.com.

In the following chapters, you will read about how exciting many of my days were. Forty years ago the content of this book would be very different. At that time, lawyers would tolerate me using the real names of people and companies. Today, that has changed, and I generally do not use the names of individuals in this book. If I do mention a name, rest assured that I got permission from that person to include them in the book. I regret that I was not able to use the names of many good and honorable people who are now deceased because I was unable to locate their family to obtain permission. There was one particular engineer, perhaps the best engineer I ever met, who died many years ago and had no family. Lawyers insisted that I not mention him by name. I regret that, Mr. H., and I am sorry to be unable to include you in particular.

I would like to express my gratitude to the many people whose names are mentioned in this book for their information and cooperation. Most of them have helped me carry the torch on parts of my journey across the golden years of aviation. Over the years, some of these people carried the torch on another part of the journey, and now the torch is carried by others.

Enjoy the stories about my time as the torchbearer.

2

The Beginning

On September 10, 1976, British Airways (BA) Flight 476, a Hawker Siddeley Trident en route from London to Istanbul, collided midair near Zagreb, Yugoslavia (now Croatia), with Inex-Adria Aviopromet Flight 550, a DC-9 en route from Split, Yugoslavia, to Cologne, West Germany. The collision was the result of a procedural error on the part of Zagreb air traffic controllers. All 176 people aboard the two aircraft were killed, making it, at that time, the world's deadliest midair collision. That was my first day at the Aero Technical University in Zagreb; my first day at school where I was supposed to learn about aerospace, safety, flying, avionics, aerodynamics, hydromechanics, and math, among other things. I realized quickly that aviation is a serious business.

My first day at Aero Technical University was an important milestone for me, because I had to earn my place there. My family is a family of watchmakers with a clock/watch repair shop established in 1892 by my great grandfather, Nikola Vardian (Figure 2.1).

FIGURE 2.1   My great grandfather, Nikola Vardian (the first watchmaker of our family).

He used to repair the tower clocks on churches in the Austrian monarchy, but eventually, he could not climb the towers anymore and established his workshop in the small town of Glina. His daughter, my grandmother, was also married to a watchmaker, Karel Jesenek, who learned his workmanship in Vienna. After his death, their daughter Helena married the watchmaker Djuro Jozic, who later became my father. With such a longstanding family enterprise, it was important to keep the business going. Therefore, my father told me: "Son, if you prove that you can repair watches and jewelry, you will have a decent profession, and after that, you can go to your 'funny university' and play with the airplanes." That is how I got myself to Aero Technical University in Zagreb, where it took me 3.5 years to finish my degree. I don't want to brag, but I was the fastest student. My graduation assignment was about cockpit clocks, just to show my father that I still cared about time and clocks.

Little Boxes

After my graduation, I moved to The Netherlands, married Connie Noordzij, and got a job at KLM Royal Dutch Airlines (Koninklijke Luchtvaart Maatschappij). This is another day

I remember well: On June 1, 1980, I started my career at a major airline. My first job was in the avionics shop, repairing little boxes. Well, I called them little boxes. I was repairing Gables and other control panels. At that time, my company was flying Fokker F27s, F28s, DC-8s, DC-9s, DC-10s, and 747 classics. There were tons of little boxes.

There was a song about little boxes, and I sang it every day:

Little boxes on the hillside little boxes made of ticky-tacky
Little boxes on the hillside
Little boxes all the same
There's a green one and a pink one
And a blue one and a yellow one
And they're all made out of ticky-tacky
And they all look just the same.

After a couple of months, I was transferred to the multiplex systems group, where there were even more little boxes. My ambition was to manage projects, but I knew that I needed more technical knowledge. Nevertheless, there was a project in the movement control department: they wanted to improve their workspace and notify all interconnections of their communication equipment. At least it was a project—a lousy project, but a project. I was glad to have an opportunity to score and show what I could do. After the project was finished successfully, I was asked to join the test equipment maintenance and calibration laboratory. So, I got a new job.

That was how I avoided going back to the little boxes. Test equipment maintenance and calibration was a different ball game. There was a lot of equipment: test sets and measuring devices to calibrate. It was not boring work at all. At that time, I encountered a problem: Not all of my education credentials earned in Yugoslavia (a communist country at that time) were recognized by the Dutch government. I decided to sign up for a degree program at the University of Amsterdam. It was an Electrical and

Telecommunication Technique program. That was a tough time. I spent the whole day working in the avionics shop and every evening at school until midnight, every day, all year long, for five years. Those were long days, my friend. But I had fun, because at work, I was learning a lot about testing aircraft Line Replaceable Units (LRUs), and school was fun because it was a very interesting program, and all the theories I was learning were relevant to my activities at work.

After I graduated from the University of Amsterdam, I found that my Dutch diploma broadened my horizons. I now had two bachelor's degrees: one from the Aero Technical University in Zagreb and one from the University of Amsterdam. It was no coincidence that my second graduation assignment involved again the clocks. I decided to build a small computer and connect a self-made receiver for a frequency of 77.5 kHz. The receiver received a signal from a very accurate atomic clock in Germany. The small computer decoded the signal and presented it on the seven-segment display. Now I was not only a watchmaker but also a telecommunication engineer who specialized in time signals.

It took me some time to get a job in the engineering department. For nine months, they kept me at my old job because they couldn't afford to lose me. My low-level managers at KLM did their best to keep me on the old job because it was better for them to have me there. They were not acting in the best interest of the company but in the interest of their own positions. I was in a similar situation many times later in my career. This is the Dutch style of management, which can be frustrating. Eventually, after a lot of negotiation, I was able to move on.

Engineering

It took me a few more months to understand how airline avionics engineering operates and what it means to be an avionics engineer. My job title was Air Data Engineer. I had to manage components and systems (altimeters, air data computers, pitot tubes, Total Air Temperature [TAT] probes, Angle of Attack [AOA] sensors, etc.). Perhaps the most important aspect of my position was that I became the engineer responsible for all Greenwich Mean Time (GMT) clocks on the KLM fleet. This fit perfectly into my family tradition, which you might also call an obsession.

The avionics shop wanted to establish a lab for the calibration of very accurate air data instruments. The company purchased two Honeywell Air Data Test System (ADT) 222 pressure standards. Those standards were not very stable and needed to be calibrated every three months. One of the devices was used, and the other was sent for calibration. By the time the first unit was back, the second one was due for calibration. The costs of outsourcing and turn around time were good enough to calculate a positive business case for the lab. The lab was supposed to be connected to the national standard. To become part of the Dutch standardization organization, the lab needed policy and procedures, had to be independent from production departments, and needed the accuracy calculation for its standard. That last component was my assignment. Before I started to investigate how to perform the calculation, I was sent to Houston to a company called Ruska for training. They were building so-called deadweight testers, and I enrolled in their course for deadweight testers and accuracy calculations. This was a good experience for me. I was able to see Houston, get a feel for Texas, and learn some new things at Ruska.

Pressure measurement was an interesting area, especially because we were at the edge of accuracy. Pressure is not just force per area, as you learn at school. When you are working in the area of extremely high accuracies, you must take into consideration that the area is dependent on temperature and that gravitational force plays a role in the equation. When I returned to my office in Amsterdam, I was able to calculate the accuracy for the KLM lab. Then I ran into my friend Bas (Bastian), a software engineer. I told him what I was doing and that I did the calculation, but I needed a cross-check. I explained that I had a formula with many variables and asked him if he could write a simulation program in SAS language (Statistical Analysis System [SAS] was used for statistics at that time) and if he would make a simulation with thousands of runs. Bas helped me out. Every parameter was randomly changed 10,000 times within its uncertainty interval, and the total uncertainty was calculated. Such statistical calculation and classic calculation didn't deviate more than 0.1%. I was happy with the result.

A few months later, we opened the pressure lab. It was a big celebration, and many KLM big shots were present. To my surprise, a gentleman from the government calibration institute mentioned my name in his speech, telling the attendees that the whole process was accelerated due to the work of Engineer Jozic. That was nice to hear. Mind you, at that time, I was 27 years old, which is quite young for an engineer, and my name was already known to many managers.

Our avionics shop was becoming self-sufficient. We were no longer dependent on logistics and external parties. ADT222 was calibrated in-house with a turn around time of just four days. That created the capacity in our shop to accommodate more work and the opportunity for some third-party calibration of accurate pressure equipment. Within the engineering department, I got the reputation of being a good problem solver. Every time there was something weird going on, they asked me to fix it. My early work at the calibration lab and my theoretical knowledge was a great combination and gave me a big advantage. I also had a lot of experience in the avionics shop because of my work in the test equipment maintenance lab. I had much more knowledge and experience than other young engineers coming straight from school, and this served me well.

At that time, KLM was flying DC-8s, DC-9s, 747 classics, and DC-10s. We had a small daughter company, NLM, which later became Cityhopper. They were flying F27s and F28s. The F27 fleet was only in the hangar during the weekend. Therefore, I didn't have much contact with that type of aircraft. Cityhoppers had their own engineering, so there was not much contact with them either, but at one point, I got a call from their head of engineering, Mr. Verbeek, who used to work at KLM Engineering. He explained that they had a problem with the GMT clock on one of their aircraft and asked if I could determine what the issue was. I called the planning department and found out that the aircraft would be at the apron at the Schiphol-Amsterdam airport the following morning. The next day, I went there to see what was going on with that GMT clock. To my surprise, in the F27 cockpit was a mechanical clock with no wires attached. My watchmakers' genes came out to play, and my engineering spirit was at its peak. I asked the technician (tech) to remove the clock and give it to me, explaining that I would return it repaired. That was the deal. As a reminder, it was the 1980s, and this kind of interaction was possible at that time. He removed the clock, and the hole in the cockpit panel was closed with duct tape to prevent the cold air from blowing through the hole onto the lap of the pilot. "Clock removed for repair" was the entry in the maintenance logbook.

That day I took the GMT clock home and opened it in my watchmaker's workplace in my attic. (Yes, I have a watchmaker's workplace—it's in my genes.) It was evident to me that the spring of the clock was broken. The following day I went to a small shop nearby that sold piece parts for watchmakers. I showed them the spring and the guy pulled a new spring from a drawer. "7.35 florin!" he said. That would be approximately three US dollars (USD). I was happy with the spring, which I installed in the clock in no time. The clock was running fine, and I was quite proud of myself. The next day I had to intercept the F27 with the duct tape on the GMT clock position at the cockpit panel. The same tech from a few days prior reinstalled the clock and placed a new entry in the aircraft maintenance log: "GMT clock repaired." My job was done: Nice and easy.

The bill for the spring was sent to the head of engineering from Cityhopper, and a few days later, I received 7.35 florin in an envelope at my desk. It was the type of interoffice correspondence envelope with holes. Younger generations do not even know that such envelopes existed in the 1980s.

This is just one of my idyllic experiences in aviation beginning in the 1980s. Such funny things were possible at that time. It might be possible now as well, I suppose, but the awareness is different. Nowadays we know that we need a decent paper trail, and we must use approved parts. You can't just go to the shop next door and buy a part and install it in an aircraft. I was an engineer, and my background (watchmaker) and sound engineering judgment helped me solve this GMT problem. As far as I can recall, this was the only time that I deviated from the standard procedure. I did it because I knew that I could fix the clock. If I hadn't been a watch-maker, I wouldn't even have bothered to try and fix it. My advice would have been to remove the clock and send it to an Original Equipment Manufacturer (OEM) for repair. Imagine how much that would have cost and how many days the aircraft would have had to fly without a GMT clock.

DADCs

I gained my engineering skills in this environment. The largest portion of my work was Service Bulletins (SBs). I was getting at least ten SBs per day. In those days, we were happy if components had a Mean Time Between Unscheduled Removal (MTBUR) of 5,000 hours. In our pool we had a lot of expensive components that were almost constantly subject to modification. One such compo-nent was the Digital Air Data Computer (DADC). On the 747 classic there were two versions of air data computer: CADC (Central Air Data Computer), which was an old analog computer, and a new version known as DADC (Digital Air Data Computer). Unfortunately, the old analog air data computer was much more reliable than the new one. We decided not to do the retrofit, but every time a new 747 arrived, we got two unreliable DADCs. Eventually, we made a deal with Bendix to modify the status of computers from −2 to −4 and −3 to −5. It was a very expensive modification, and Boeing was paying for it. The precondition was to send those computers to the Bendix shop at Los Angeles International Airport (LAX) for modification. The modification program was running for many years, and everybody was complaining about it. I wanted to buy one additional spare DADC to use for modification purposes, but my management refused to spend 50K USD on the computer. They complained that they already had to purchase such a computer because one computer was missing, and they had to replace the missing computer. Because of this, the modification program was running slowly.

At the end of the 1980s, KLM was phasing out DC-10s and buying McDonnell Douglas MD-11s. DC-10s were mostly flying to Africa. After the last airplane left, we got back all the spare parts, which were stored in outstations in Africa. One 747 DADC came back from Lagos (LAG), and the pieces of the puzzle came together. When I was running that modification, one computer was reported missing. It turned out that this was because somebody from the logistics department made a mistake and sent the computer to LAG instead of LAX. The computer arrived at LAG, and local staff thought that it was a spare part for DC-10s that were coming to Lagos every day. The computer was placed on the shelf and never used, because it was not a DC-10 part number (p/n). Nobody ever asked for it, nor did they ask why a 747 p/n was on the shelf. The guy in Amsterdam was not aware of it, and he just inserted the letters MI (Missing) into the computer system. Another department was responsible for maintaining the spare stock, and they ordered a new computer.

Five years later, we had overstock in the 747 pool because the missing computer was found again.

This explains why many people working in airline maintenance departments joke that there is a wormhole in every maintenance department. Aircraft parts just mysteriously vanish. A few days, weeks, or sometimes years later the missing part reappears somewhere. We had one person in the logistics department whose job was to walk around the technical area and search for missing parts. He located most of the wormholes within the technical department and saved the company a lot of money by tracking down the missing units.

Such was the beginning of my glory days.

3

ASDAR: A Weather Eye in the Sky

Back in the 1970s, an interesting scientific project called ASDAR, or Aircraft-to-Satellite Data Relay, was developed. At that time, communication systems such as Aircraft Communication Addressing and Reporting System (ACARS), Satellite Communications (SATCOM), High Frequency (HF) data link, and Very High Frequency (VHF) data link were not mature and available for use. The National Aeronautics and Space Administration (NASA) was asked by WMO (World Meteorological Organization) to develop ASDAR—an airborne data collection system that gathers meteorological data from existing aircraft instrumentation and relays it to ground users via a geosynchronous meteorological satellite. Following are details on the beginning of the ASDAR project. The results of the first test flight on a commercial 747 aircraft are also described. On the Internet you might be able to find the flight report issued by NASA in 1977: This was the first version of ASDAR. Pan American Airways (PAA) installed it on a 747 and received the Federal Aviation Administration (FAA) STC (Supplemental Type Certificate).

The report says: "The ASDAR test flight is a success, and the program will continue as scheduled with a one-year evaluation period on PAA aircraft NP657 PA."

FIGURE 3.1   ASDAR (Aircraft-to-Satellite Data Relay), the one-way transmission direction.

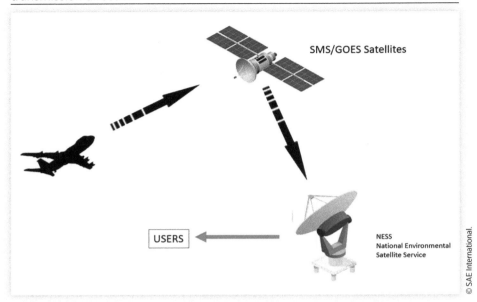

As can be seen, the aircraft system sends the obtained data via satellite to the ground station at the NASA facility at Wallops Island (Delaware). From there, it is sent to NESS (the National Environmental Satellite Service) in Maryland. After repackaging, it is sent to the users, which are meteorological organizations.

Meteorological offices worldwide needed accurate and cost-effective weather stations. They were using expensive ships, and they were not happy with the limited amount of data obtained from the lower atmosphere (Figure 3.2).

FIGURE 3.2   ASDAR data sampling profile during the flight.

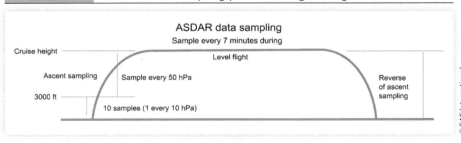

Following the PAA STC in 1977, many discussions took place. Some airlines followed the PAA model and installed the first ASDAR version in their 747 aircraft. My airline did this. The system was also installed in a KLM 747 aircraft PH-BUB, but not for long. After some time (two years), it was decided to deactivate the system. Eventually, NASA later asked the contributing airlines to remove the LRUs from the aircraft and return them to NASA. The units were removed in the mid-1980s. The ASDAR units were not owned by the airlines but by the World Meteorological Organization (WMO), which was behind the project. One of my fellow engineers issued the engineering order and removed ASDAR LRUs and sent them to NASA. This was the situation until the end of the 1980s. Then WMO contracted the British company Matra Marconi Space to build a new ASDAR system. The new system design was lighter and more reliable—meant to be a fit-and-forget system.

System Architecture

The design of the system was simple and effective. It was a two-box system with a personality module and flat antenna. The ASDAR was connected to the standard aircraft systems, and no scheduled maintenance was required. The system consisted of Data Processor Units (DPUs) and a Satellite Transmitter Unit (STU). The DPU was connected to aircraft systems and collected meteorological data (altitude, wind speed, temperature, wind direction, aircraft position, etc.). Every seven minutes en route, a snapshot of data was collected and stored. Every hour, the satellite transmitter was activated, and the data package was sent to a meteorological satellite. Because there was no active synchronization of ASDAR and the satellite clock, a very accurate clock was required in the ASDAR DPU. The clock was set to the correct GMT.

During a climb, the snapshot of data was not collected as a function of time but rather as a function of pressure. This gave the meteorological people a good idea of the weather profile around the airport.

One feature in the ASDAR system was never used. The system was capable of executing an emergency transmission if triggered by a crew member. To enable this feature, a small push button had to be installed in the aircraft. After activation (pushing the button), a preprogrammed message is sent to the ground station. This was designed for use in case of aircraft hijacking. Aircraft were not yet equipped with SATCOM, and the ASDAR message via satellite could be very effective and fast. The hope was that, after 1992, the ASDAR system would be installed on approximately 25 aircraft of different operators. The system was reliable, and no problems were expected during operation.

Several airlines lobbied for the installation of the new ASDAR in their aircraft. The seven Member States of WMO, including The Netherlands, managed the operational implementation of ASDAR. A closely linked network of carefully chosen airlines and flight routes was established. Specific attention was given to sparse data areas such as oceans and deserted landmasses. Our KLM aircraft flying to the Caribbean were supposed to take measurements regularly and provide the data to aid in the prediction of hurricanes. These data complemented the data from the radio probe network in the southern US.

KNMI

The Dutch Meteorological (Met) Office, or the Royal Dutch Meteorological Institute (KNMI), along with the meteorological offices of six other countries, was pushing for the installation of ASDAR on our 747s. My airline was a national airline, and we had to work on this project whether we agreed with it or not. Unfortunately, there was an internal problem within the company. Many experienced engineers were retiring, and the new staff didn't have a clue about how to do the STC (Supplemental Type Certificate) to get a new system approved for the 747 aircraft. Company management decided to keep discussions going hoping that the Met Office would forget about the project. But it did not work out that way. One day my boss came to me and said: "Here is some information, go to the meeting with KNMI and make something of it. We can't escape it anymore."

That is how we started negotiations with KNMI. Eventually, they were very fair and even paid the fee for carrying the payload on the aircraft. Soon the contract was signed, and I got a new project. I considered it good luck: When nobody was familiar with the STC procedure, I had a chance to be a pioneer. The project was not operationally risky, and I could learn all the steps in the STC process. I was at the beginning of my engineering career, and I used the know-how gained from this project for the next 30 years.

ASDAR Implementation

My first business trip for the project was to the United Kingdom (UK) to the manufacturer of ASDAR. As always, airlines had to save money, and I made the trip alone. It was a wonderful experience. For the first time in my career, there was a driver holding a sign with the name "Jozic" written on it. He took me to the Marconi plant in his big Ford Scorpio. There I met fellow engineers who were working on the ASDAR project. We had four hours of discussions and a review of their documents. By the end of the day, I was convinced that I could do the work easily, and they would support me. There were a few stand-out items that were essential to the project. The first was antenna installation. This was new for me because it required sheet metal work, and I was an avionics guy. My plan was to get a fellow structural engineer to help me. Next was the 42 ft long Heliax cable, which was important due to the high frequency of the ASDAR signal.

The cable was a copper tube, and it needed special care. The length was critical, and installing the connectors and clamps was a very complicated process. The next essential item was the cesium clock in the unit. It had to be synchronized with the atomic clock in Mainflingen. With this new knowledge and a fairly solid plan, I went back to Heathrow in the same Ford Scorpio.

Upon my return, I started working on an engineering order. It was not that complicated. I got the old NASA antenna installation drawings from Marconi with quite good instructions for wiring. I was able to trace the required signals from the Burndy blocks behind flight data acquisition units. I also issued a few drawings with instructions on how to install the Heliax cable and how to install those big Heliax connectors. My choice for aircraft tail number was PH-BUM. That became my favorite 747 because I had the opportunity to use that tail number for the first installation of many other systems (TCAS, SATCOM, Enhanced Ground Proximity Warning System [EGPWS], elementary surveillance/enhanced surveillance [ELS/EHS], glass cockpit, etc.). ASDAR was first.

One day in the office, I bumped into my friend, an engineer named Bas (Bastian)—yes, the same engineer who helped me with the calibration lab. Bas was a software engineer from the Aircraft Data Management group. We talked about our work, and I told him about ASDAR. "ASDAR, ASDAR it sounds like a dog name," he said. And continued: "All that data that you are collecting we can also collect on 767 and 747-400 and send to the ground station via ACARS link." That was interesting information, and I decided to share it with KNMI. They were very interested in ACARS, and I invited Bas to explain it during KNMI's next visit to my office. They wanted all the data we had, and we could make it available not only via one ASDAR system but via 39 additional aircraft. The data from 39 aircraft were not directly available but had a delay of a maximum of four hours. We didn't have SATCOM in the aircraft, and the ACARS transition had to wait until the aircraft was in VHF coverage. Bas also got a project: He was now selling data to KNMI. A lot of data!

The installation date was approaching, and I had sent a fax to KNMI that we were ready to go. Yes, back then we often used fax machines to communicate. To my surprise, they asked if I could contract a television (TV) crew to record the activity. OK, I did that too. KNMI wanted to have some video data for their internal use, so we had the 747, the techs available, and the TV crew. We were ready to rock and roll.

I knew that there would be some unexpected problems, but I am an engineer, and engineers are on Earth to solve problems. Of course, the problem was reported in the structural provisions of the antenna. There were some minor deviations in the drawing compared to the actual situation in the aircraft. The brackets didn't fit smoothly between stringers. The tech trimmed it a bit, and the problem was fixed. Wiring installation was smooth with the exception of a few wires that had to be spliced. I am not in favor of using splices, but the wiring tech told me: "A good splice is better than a piece of wire." I had to redline the drawings and submit the changes to the Designated Engineering Representative (DER) for approval. Approval went smoothly, and the installation was done within 24 hours. Not bad, not bad at all.

The modification process was running and also waiting. We were done, but now came the final test. For that test the aircraft had to be outside, and the antenna was supposed to "see" the satellite. But the 747 was in the hangar and the tow truck was still on the way. We had to wait, wait, and wait some more. After two hours of waiting, the tow truck arrived and pulled PH-BUM from the hangar to the platform.

We had power on the system, and I went down to the avionics bay to initiate the transmission. Through the keyboard at the front side of the DPU, I first had to synchronize the clock and then insert the code to trigger the transmission. I initiated the transmission three times but had no idea if the transmission was successful. The only way to find out was to go to the office and call Matra Marconi Space in the UK and ask. In 1992, when we were doing that STC, we didn't have mobile phones or SATCOM on board, so I had to walk to the office to make the phone call. Eventually, I got the OEM on the line, and they confirmed receipt of the "ready" message three times. Great! My project was successfully completed.

The same evening the aircraft was ready to go, KNMI surprised me. On the evening news, one of the weathermen mentioned my ASDAR installation on national TV. It was a nice surprise and quite unexpected. I was only able to take a unsharp picture with my fotocamera of my TV screen showing that historic moment.

The ASDAR system was preprogrammed in such a way that the data set was transmitted in a certain time window. For my aircraft PH-BUM was assigned minute #26 of each hour to execute the transmission. Therefore, the cesium clock in ASDAR had to be synchronized with the atomic clock located in Mainflingen, Germany. If the clock drifts for any reason, the aircraft will transmit in a different minute and disturb the transmission of another aircraft whose transmission was assigned in minute #27 or #25. I was proud of my installation and was certain that PH-BUM would transmit its data every hour in the 26th minute. In later years, only once did I have to go to the aircraft to synchronize the internal cesium clock with the atomic clock called DCF77 in Mainflingen. After the ASDAR clock was synchronized, I could sleep well.

DCF77

DCF77 is a German long-wave time signal and standard-frequency radio station. It started service as a standard-frequency station on January 1, 1959. In June 1973, date and time information was added. Its primary and backup transmitters are located at 50°0'56"N 9°00'39"E in Mainflingen, about 25 km southeast of Frankfurt am Main, Germany. The transmitter generates a nominal power of 50 kW, of which about 30 to 35 kW can be radiated via a T-antenna at a frequency of 77.5 kHz. DCF77 is controlled by the *Physikalisch-Technische Bundesanstalt* (PTB), Germany's national physics laboratory, and transmits in continuous operation (24 hours). The transmitting antennas are visible from the highway. If you pass Frankfurt Ring driving northbound and take the exit to Sauerland highway, you can see the large antennas immediately to the left of the highway.

I passed by that area many times, and I didn't know that those were antennas of the most important clock transmitter in Europe: DCF77 (Figure 3.3). All European airports and train stations and all official clocks are synchronized with DCF77. Just a few months ago, when I saw the picture on Wikipedia, I realized that I pass by those antennas at least twice a year. I can say: been there, seen that … many times.

FIGURE 3.3 Left and Right: DCF77 antennas in Mainflingen near Frankfurt.

© PTB. © PTB.

ASDAR Spin-Off

A software package simulating ASDAR can easily be implemented into the Application Control and Management System (ACMS) software of modern aircraft. For data transfer, ACARS message via VHF, HF data link and SATCOM can be used. In close cooperation between KNMI and my airline, 39 aircraft became available with the new generation of aircraft observation systems. Currently, nearly 100 aircraft from many airlines are providing observations (temperature, wind, and air pressure) every hour. These observations are made at flight level and during ascent and descent, and are available in near real-time to KNMI and other national meteorological organizations on a global scale. The ASDAR system itself was decommissioned and removed from the aircraft in 2003 before the aircraft with tail number PH-BUM was sold to Holiday Airlines in Bangkok.

4

TCAS Pete

The first recorded midair collision was in Italy in 1910, when a French pilot collided with a British pilot. Both were flying in airplanes built of wood and sealed fabric skin over the airframe. Fortunately, both pilots survived. The British pilot never flew again, maybe because of injuries or perhaps because of a new fear of flying. The first *fatal* midair collision occurred in 1912, when two French pilots crashed into one another in an early morning haze. Lieut. Peignan was taken dead from among the debris. Capt. Dubois died within an hour. After that, there were hundreds of midair collisions. It became so dangerous that engineers were called on to solve the problem and save lives.

Prior to the 1980s there was no system in the aircraft to warn pilots about approaching traffic in their aerospace. Only ATC (Air Traffic Control) could predict and avoid possible collisions if there was radar coverage. With no radar coverage ATC was dependent on a clock and the position was reported by the pilot. It was a pretty scary situation.

In 1981, the FAA announced a decision to implement an aircraft collision avoidance concept called the Traffic Alert and Collision Avoidance System (TCAS). TCAS is an aircraft system designed to reduce the incidence of midair collisions between aircraft. It monitors the airspace around an aircraft for other aircraft equipped with a corresponding active transponder. If there is no transponder, the aircraft is invisible to TCAS. TCAS is independent of Air Traffic Control (ATC) and provides information to the pilot about the presence of other transponder-equipped aircraft that may be in the vicinity and present a threat of collision on the display. When the FAA mandated TCAS, it was the start of a whole new aviation era.

The TCAS Era

Most airlines didn't know much about TCAS—I will keep calling it TCAS, although in literature you can also find the acronym ACAS, which stands for Aircraft Collision Avoidance System; we are talking about the same beast. In the beginning, it was called TCAS II. On Wikipedia, there is an in-depth article about the evolution of TCAS and the differences between TCAS I and TCAS II, as well as TCAS III and TCAS IV. I won't go into those details here. TCAS I was a primitive system used in general aviation and never actually used on a large scale. Airlines eventually implemented a more mature TCAS II system.

Because the airlines didn't know much about TCAS, OEMs organized conferences about their systems. There were three major players: Honeywell, Rockwell Collins, and Allied Signal. Each held a conference to teach people about TCAS and to connect with potential customers. I was busy working on other projects, and my management decided to hire a guy named Piet van den Berg (later known as TCAS Pete) and made him responsible for the TCAS project. He attended all TCAS conferences and workshops and was working toward the implementation of the system for our company. We had various aircraft types, and every type had to be equipped with the TCAS system. It was not an easy task. Piet's desk was not far from mine, and I could follow all of his experiences and concerns.

Project Phases

A project such as this has a few phases: Feasibility Study, Specification, Preparation (including business case and financials), and Accomplishment and Evaluation.

The first phase required a decision on possible system architecture. It was essential to know what could be installed in our aircraft types. Here we were constrained by the typical Dutch mentality.

The requirement was not to provide a straight and elegant solution but to be cheap. For the rest of the book, I will try not to say *cheap*, but rather *cost-effective*.

The easiest way to implement the system was to go to one OEM and buy the system and certification. But no, management complicated the process for us engineers. The TCAS computer came from OEM A, transponders from OEM B, installation kit from OEM C, and certification from OEM D. That might be the most cost-effective solution, but the engineer had to force all parties to work together and deliver the functioning system. Well, my buddy TCAS Pete was in a difficult position. Most engineers were responsible for a few systems and did some projects in between, but Pete had a full-time job with TCAS. At some point, he became known around the office as "TCAS Pete." In the next phases of the project—specification and preparation—TCAS Pete got some production and preparation engineers to work with him. For the TCAS system in our fleet, the situation was as follows: On a 747 classic, the choice was a Honeywell TCAS computer and Rockwell Collins Mode S transponders. The TCAS VSI (Vertical Speed Indicator) was also from Honeywell, as were the antennas. The Mode S-TCAS control panel was from the manufacturer Gables.

Early on, the market for these products was not developed, and for the first few 747 classics, we had to assemble the kit in-house. This was not easy. The basic document was the Honeywell STC data package, which was complicated. It included an antenna switching by means of coax relays, which was critical because the coax cables had to be built very precisely.

At one of the conferences, TCAS Pete discovered a small company from Milwaukee that specialized in cables. After working on our third aircraft, we started to purchase their coax cable kit. The owner of that small company had a contract with a local hospital to build a few hundred extension cables for hospital equipment. He was just starting his company and was working alone. While assembling cables in his kitchen with the TV switched on, he heard a reporter mention that the aviation community was hosting a conference about TCAS in Milwaukee. It looked like there would be a lot of work for the aviation industry for providers of cables for sophisticated computers. It was an intriguing story, and since the conference was in Milwaukee (his hometown), the owner decided to check it out. During the conference, he had an epiphany. He decided to build a coax cable kit for TCAS installations. This was quite an undertaking, because he had to obtain all the required approvals and organize the production workshop to build the cables. The installation kit consisted of two sets of Mode S transponder cables for the top and bottom ATC antenna and eight TCAS cables, four for the TCAS top antenna and four for the TCAS bottom antenna. That is how he started in the aviation business. Later, he grew his one-man shop into a multimillion-dollar company, which became a major player in the production of aircraft installation kits. His story is truly the American Dream.

We were very happy that TCAS Pete found this company at the conference. Those installation kits made our lives much easier. A few months later, we planned the first installation in one of our aircraft. We decided to do the STC, and the Honeywell team came to us to help with the certification (Figure 4.1).

FIGURE 4.1   Engineers at 747 upper deck flying back to Amsterdam after finishing the TCAS test flight program.

Courtesy of Peter Berg.

By that time, we had a good team of techs who were very adept at installing wiring and equipment. There were no big obstacles, and, honestly, Pete had done a great job in designing the system, getting all the necessary parts, and coordinating with Honeywell and other parties involved. As far as I remember, there was only one glitch, and that was the antenna switching. Actually, it was *not* switching, and that was the problem. The issue was in the interface between the ATC/TCAS control panel and coax relays. For antenna switching, two diodes had to be installed to pull down the switching signal to zero. Otherwise, the relays would not trip. Our friends from Gables issued the SB for the control panel, and after we accomplished the SB, the system worked as advertised. All went smoothly. In later years, Gables saved our skin a few times on different installations. It was great to have such friends at Gables in Florida.

We thought everything was under control and that all the following installations would be a piece of cake—and they were. But our sales were not sitting still. We had a great number of customers worldwide because we were good at 747 maintenance. Half of our hangar work was for third party customers, and those third-party customers had to install TCAS as well, because it was a global requirement. We were forced to share the work among the avionics engineers to serve all our customers. We were lucky that we had TCAS Pete among us because we could ask him for advice and discuss the issues. TCAS Pete did not keep his know-how to himself: He was always willing to help.

I was responsible for our customer Garuda, who decided to purchase the system and components from Allied Signal (previously known as Bendix). From their DER, I got the TCAS installation manual, some advice, and their telephone number. That was at our first meeting. Afterwards, I had to develop the system and send them my drawings for approval. During this process, we had telephone calls almost every afternoon. We didn't have email and all communication was via phone or fax. The installation was made easier because Garuda decided to purchase everything from one OEM. The DER had sent me one additional TCAS computer just in case we needed to replace the original one if it failed during testing. That computer arrived at our incoming goods inspection and after that was lost. The aircraft installation went very well, and after the test flight, our 747-test pilot came to me, shook my hand, and congratulated me for a job well done. The next four aircraft were business as usual. All four aircraft were in our hangar for D check, meaning that we had a customer for 20 weeks in our facility.

After the fourth aircraft left, I got a telephone call from our engine shop. The guy on the other end was complaining that there was a heavy box in their corridor with my name on it that they were constantly tripping over. I went to the engine shop and found the missing TCAS computer. It had been sent by Allied Signal to my attention. That is how I learned that, for every package, I had to issue a purchase order (p/o). If it was sent without a p/o, it would almost certainly be lost. I was lucky the computer was found, because a TCAS computer at that time was very expensive (70K USD).

Eventually, we began working with a third major manufacturer. Our customer Avianca from Colombia decided to purchase TCAS from Rockwell Collins, and we had to design the installation and complete the certification. At the same time, MEA (Middle East Airlines) came to me with their 747 classic. They wanted us to support the installation in Beirut.

We had a great team, and I was playing with the idea of approaching our management with a proposal of establishing a dedicated modification team. My staff in the hangar were enthusiastic. My colleague engineers were also willing to step into that business. It was much more rewarding to do the hands-on projects than sitting in the office reading SBs and investigating whether the accomplishment of the SB would be an improvement for the fleet.

My idea was to have a team of 15 experienced techs who could accomplish every aircraft installation. The team could be supplemented by an additional five to ten people from D or C check as needed. We had a lot of requests for installation, and the team could go to the customer site and do the installation there. My management didn't want to hear such ideas. They supported ad hoc installations for customers, but there was no interest in starting a modification team.

Lebanon

I had to go to Beirut to discuss the installation for the MEA project. The war in Lebanon had just ended, and I didn't want to stay there for a long time. I arrived via Rome in the morning, had the meeting, and took a night flight back to Amsterdam. The meeting was promising, and the people I met were very friendly. We decided to do the installation in Beirut. They had an FAA repair station and were able to do the wiring. I could send two licensed 747 techs for the whole period of installation. For the last three days of installation, I had to return to Beirut to sign the 337 forms for the release of the aircraft. For this purpose, I got an FAA authorization.

To be on the safe side, we told the authorities in Lebanon that we were only supervising the installation and not working there. We were in a relatively safe area at the airport, and our hotel, Summerland, was quite safe because of a very big and strong fence around it. The buildings around and in the center of Beirut were mostly destroyed. Everything was in bad shape. The local team brought us to a restaurant on the other side of Beirut. There was one road crossing the city center, and in all the streets, right and left, there were ruins of buildings and rubble, concrete blocks and bricks three stories high. It was a frightening experience in Lebanon after the war, but the installation was a success, and the customer was happy.

Columbia

It was a different story at Avianca in Bogota, Colombia. They wanted six techs for local support, and our sales team agreed and made a deal. My best people went on the mission in Bogota. They had already done 25, perhaps 30 installations on 747s, and there were no risks even with the installation of the Rockwell Collins system. My problem was that we had a bad telephone connection, and I was not able to talk to the team every day. One day, the local Avianca engineer called in a panic and told us that our whole team had been arrested. This set off a major panic in the office. Six people in prison in Colombia is bad news. Here is what we heard: My tech buddy and his team were in the 747. Avianca techs were installing wiring, and my buddy and the team were assisting. At one point, the local staff left the airplane and the police came on board.

The police accused our team of illegally working on the airplane and arrested them. We were notified a few hours later, and our sales and lawyers started negotiations. Fortunately, we had a staff member who spoke Spanish, and that made it easier to deal with the local police in Columbia. In the end, we had to pay 5,000 USD to the police. After they received the cash, they released my technicians. The technicians were allowed to finish the installation and, after that, were required to depart within 24 hours. What a relief! The lesson learned for me was not to send my techs on any mission in a corrupt country. The Colombian chief of police got a 5,000 USD bonus that Christmas. The installation didn't bring the tech team any additional revenue, and was a frightening experience.

The TCAS Seed

TCAS Pete planted a seed. We became a very experienced engineering organization, and we had great techs. The sky was the limit! Boeing 747s were our babies, and we loved our jobs. TCAS changed the world of aviation. Many new companies entered the aviation ecosystem. Soon, there were a lot of companies that could provide the kit, installation package, and even the certification. We all learned from the process, and this knowledge was later used for SATCOM installations, Terrain Avoidance and Warning System (TAWS), cabin systems, etc.

TCAS was probably the most important project at the engineering department in the years between 1980 and 2000. It was big, complicated, and very extensive. Many people were involved, and the entire aviation community learned so much that, in the next 20 years, all that knowledge could be used to benefit other projects. The FAA mandate was dated January 1, 1993. All commercial turbine-powered transport aircraft with more than 30 passenger seats (or Maximum Takeoff Mass [MTOM] above 33,000 lb or 15,000 kg) had to be compliant on that date. Before this mandate, the industry had to learn to design LRUs and systems, connect that all together, train pilots, be able to repair components, do the testing and accomplish STCs, and deal with authorities, lessors, sales purchasing, and many other things. These days, nobody cares about TCAS: It is the standard equipment on board the aircraft, installed by airframers, and it is always working. No one remembers that we engineers saved many lives by implementing such a system. In Chapter 2, I described the midair collision between Adria DC-9 and BA Trident on September 10, 1976. If those airplanes had TCAS, those 176 people would not have been killed. I am thankful that I was part of an engineering force that designed the wiring installation and worked on the TCAS project. My fellow engineer TCAS Pete did a great job of introducing TCAS to the whole group of engineers.

On a global level, our contribution was not great (approximately 200 aircraft were modified), but we managed to design the wiring for installation on at least six aircraft types based on three different vendors. These days you no longer hear about big midair collisions, and I can proudly say that I contributed my engineering skills to this increase in airline safety.

The Last 747 Classic

I n the early 1990s, people were complaining about the "crisis" in aviation. Sometimes airline companies get into problems, and the discussion around the problems references a crisis. This is not always a global crisis, and it may not be a crisis at all. It is sometimes something local caused by mismanagement or when a company is unable to change and adjust to a new situation. Deregulation was popular in the US, and some European Union (EU) companies were separating themselves from governments. Before this time, it was normal for governments to be involved in the airline business. It was a method of promotion for the country, and because airplanes operated globally, many people viewed airplanes as national ambassadors. Therefore, the airlines had names related to the countries they belonged to: British Airways (BA), Alitalia, SAS (Scandinavian Airlines System), Austrian Airlines, Yugoslav Air Transport, Swissair, and KLM (Royal Dutch Airlines). Those were the flag carriers.

There was some pressure from the US to separate the companies from governments to create fair competition and level the playing field. US companies had to run as businesses, and if they failed, they went bankrupt. Consider Eastern Airlines, a large operator based in Florida with 500 or more aircraft. The airline went bankrupt one day and is gone forever. A similar thing happened to the very respected and progressive PAA. On the other side of the Atlantic, there was Air France, always in financial difficulty, and the French government pumped millions of dollars to help them survive every crisis, which was happening every couple of years. No wonder US operators were complaining about unfair competition. Even EU operators such as BA and KLM complained: These airlines were separated from the government much earlier than other airlines, and they didn't like seeing competitors burning through a lot of cash, and then, when problems occurred, these competitors received millions from their governments to keep them in business.

Independent airlines also complained about unfair competition, knowing that Air France would receive financial assistance from the French government whenever needed. For Air France this was normal. Government support also explained why the airline had more employees and those employees had higher salaries. Air France had a very extensive and expensive organization. In such an environment, they had no qualms about ordering new aircraft. If business became bad, Air France expected (and received) government assistance. In the early 1990s, Air France decided to order a 747 full freighter. It was to be the very last 747 classic ever built, because Boeing had already started building 747-400s. This freighter with the serial number (s/n) 25266 was built between two 747-400s, last in the line of 747 classics. The official model name was 747-228F.

At Boeing, the process of building the aircraft is lengthy. It takes time to sort out the paperwork, get all the necessary parts, and build the aircraft. Air France, again in financial trouble and on the edge of bankruptcy, canceled their order. Boeing was already at the point of no return, and they had to finish the aircraft. Or did they? Boeing likely approached some cargo operators to ask if they wanted to buy a 747 freighter at a "good" price, which is how the Dutch company Martin Air purchased the aircraft intended for Air France. It was a great opportunity to get a new aircraft for a reduced price and Martin Air stepped into the deal. The Dutch are always trying to get the best bargain price, and often they succeed. This may be why Boeing pushed the 747 classic between two 747-400s on the production line. Martin Air didn't change anything in the aircraft configuration, because it was too late for last-minute changes. Boeing didn't even paint the aircraft in Martin Air livery: The aircraft was painted and delivered in Air France colors.

Company Specifications

I need to provide some crucial information. At that time, almost every company or group of companies had its own specification of aircraft. This was a legacy of the pre-1980s period. In Europe, there were two major 747 configurations. One was Atlas configuration with the representation of Air France and Lufthansa. The second big group was KSSU (KLM, SAS, Swissair, Union de Transports Aériens [UTA]). Among the configurations, there were extensive variations in cockpit design and instrumentation. The biggest difference was that Atlas had a so-called common cockpit, and KSSU had a split cockpit. This has to do with instrument switching. The difference in cockpit design introduced wiring differences and procedure differences in operating the aircraft. One of the "normal" things during that time was that almost every operator had a different Autopilot Mode Select Panel. This was not just a difference between dual and triple autopilot, but rather substantial differences in the position of switches at the panel. Boeing certified at least 40 different Mode Select Panels. In addition, there were differences in the Attitude Director Indicator (ADI), altimeters, airspeed indicators, GMT clocks, and so on.

One thing that Atlas and KSSU had in common was the switch convention. There were two conventions: Switches on the control panels of a cockpit ceiling panel could be switched "Forward ON" or "Forward OFF." It was lucky that 747 s/n 25266 had the same switch convention as the KSSU fleet; otherwise, all of the control panels would have had to be changed.

P/Ns

The wide variety in part numbers (p/ns) made work difficult for the avionics engineer. One of the most common questions for an engineer came from the operations control center. "If our aircraft was at the outstation with failure in any of the components, could we borrow it from a colleague's airline?" Due to the variation in p/ns, they always asked an engineer if they could use a different part or dash number.

And they expected—and needed—a rapid response. If the question was asked during office hours, the engineer could check it in the documentation. If the question was asked during the night, the engineer had to know the p/ns by heart. If there was any doubt, one had to drive to the office and check the documentation.

Once I was awakened at 2 o'clock in the morning with apologies. The caller gave me one minute to wake up completely and then told me that there was a 747 in Dubai with a broken standby horizon. The horizon was a no-go instrument. All passengers were already in the aircraft, and they had a standby horizon with dash number M1, which was different from the p/n on the aircraft. The original standby horizon had no "M1." The captain wanted to know what M1 meant in the part number of the indicator; without conformation that the part was okay, he would not depart. Conveniently, just a few weeks earlier, I had a SB in my hand that introduced different colors of the flag. That was called the modification M1. I gave that explanation and everybody was happy, and the aircraft departed without further delay.

This illustrates the differences in aircraft configuration. Airplane serial number 25266 was originally built for Air France in Atlas configuration and now was going to fly for Martinair in KSSU configuration. There were two reasons for maintaining KSSU configuration. One reason is pilot training. KLM did all pilot training for Martinair on KLM flight simulators, and the aircraft and simulator should match. The second reason was components. Martinair was able to use the KLM pool of components, which saved them a lot of cash on spares. The task for engineers was to modify the brand-new aircraft, changing it so that Martinair pilots could fly it and that pool of components could be used, with some minor exceptions.

Martinair Modifications

A big boss from Martinair (Martin, of course) came to our engineering group for help. Mind you, he did not go to Boeing but to KLM engineering. From the beginning, it was evident that this would be a complex and unusual job. It was a remarkable situation, and we had to take extraordinary measures. After some consideration, we forged a plan. Our representative in Seattle, Ben, had just retired, and we hired him back because he knew the people we needed to work with. That was a smart move. Ben helped us hire six Boeing engineers from Wichita, Kansas. Boeing had a facility in Wichita where they built fuselages for their aircraft and then transported them by train to Seattle. The hangars where Boeing did heavy maintenance checks and big projects like conversion to freighters or conversion to tankers were also in Wichita. Additionally, there was a large group of engineers capable of doing wiring modifications in Wichita. So six Boeing engineers from Wichita and ten KLM engineers formed a team.

Even before the aircraft was delivered, the team started to work on modifications. In total, we had 120 modification orders, hundreds of pages of changes in wiring diagrams, and thousands of parts on order. We just had to send a fax to Ben with the p/n, and the part would be ordered. Nobody ever asked how much the component cost would be. That was a fun part of this engineering job: Not thinking about costs, just being able to do our engineering work without constraint. We also received the parts very quickly, which was amazing. Here is the scope of the job:

- Common cockpit to be changed to split cockpit (K-box; M-box).
- VHF NAV (Very-High Frequency Navigation Equipment) to be changed from Collins to Bendix.
- Replace VHF communication Transceiver from Bendix to Collins.
- TCAS to be installed.
- Radio altimeter to be changed from Bendix to Thales.
- Radio altimeter indicators to be replaced.
- ADI and Horizontal Situation Indicators (HSIs) to be changed.
- Various panels in the cockpit should be changed to make these KLM look-alikes.
- GMT clocks to be changed.

- Install ACARS.
- Replace weather radar.
- Replace transponders.
- Replace standby horizon.
- Replace Inertial Navigation Units (INUs).
- Circuit Breaker (CB) panel to be reconfigured.
- Performance Management System (PMS) to be installed.
- Egg timer to be installed.

Yes, you read that correctly: an egg timer. An egg timer is just a mechanical timer. The pilots used it quite often. For example, when they fly above an ocean, they must report their position to the ATC regularly because there is no radar coverage. During the crossing, they set an egg timer for a certain number of minutes, and when it rings, they contact the ATC. This freed them up to do other tasks.

The egg timer I used was built by an old Swissair captain from Zurich who retired as a Swissair DC-10 pilot. After retiring, he opened a small egg timer factory in the basement of his house. He, of course, certified his mechanical egg timers. Various KSSU aircraft types were using his egg timers. The captain ran his egg timer business for many years, until he fell ill and could no longer build them. The funny thing was that there were two p/ns: -01 and -02. One had a black dial and white numbers, and the other had a white dial with black numbers. There was just the one egg timer in the cockpit, and we had no pilot complaints until we heard from one of our 747-400 captains. He was flying 747s, and because most of the time he was flying during the night, he insisted on a white dial with black numbers. He always complained if he entered the cockpit and found a timer with a black dial.

I hope I have shown that such a mundane thing as an egg timer can be an important cockpit tool.

We had a very impressive list of modifications to complete for Martinair. The process shows how skilled the group of engineers was. The only thing that was not in the scope of the project was the installation of a third autopilot. The aircraft has a dual autopilot, and it remained like that until the end of the aircraft's life.

We had to develop ground and flight test procedures, issue hundreds of job cards, and design appropriate project planning. For the first time, I had to compile the Electromagnetic Interference (EMI) test, though I didn't have the slightest idea how to do it. I was regularly attending sports classes at our gym association at that time, and one of the people at the gym was a test engineer from Fokker. One evening, we had a chat, and he explained the philosophy of EMI testing. It was simple. That knowledge followed me throughout most of my career. For all future STCs that we did, I was the go-to guy when it was time to write the EMI test procedure. This was not because I was so good, but because others were too lazy to learn how to do it, and it was much easier to ask me.

All the modification tasks required four months of work for perhaps 50 people. The group involved was getting larger. After hard engineering and preparation work, we could hardly wait for the aircraft to arrive in Amsterdam. It was the summer of 1991. Boeing staff were working hard on the project every day, and in the evenings they went to Zandvoort, a small town on the beach. Topless sunbathing was very popular and was something they didn't have in Wichita. The team from Kansas spent afternoons and evenings at the beach bars enjoying the view.

It was extremely rewarding when we formed a well-working team. We grew closer and became friends. We helped one another on the project and celebrated birthdays and anniversaries together.

A Celebrated Mistake

One of my Boeing buddies told me that another Boeing engineer on the team had made a mistake in wiring a few years prior to the Martinair project. The engineer was really very upset by his error. He screamed: "Oh my god I made a mistake. Terrible that I made a mistake. I am so ashamed." After that, every year, his team would tease him and celebrate the anniversary of his mistake. The team decided to celebrate the anniversary of Pete's mistake again while in Amsterdam. I told them to keep it secret and did some work in the background. I ordered a plank and installed two wires and two splices. The red wire was connected to the blue wire via the splice and the blue to the red via another splice. The inscription said, "5th Anniversary of Pete's mistake."

Of course, we said nothing to Pete. One afternoon, I went to the office, climbed on a chair, and said: "May I have your attention please." Then I told the story about a mistake that was made in Wichita that was so well known, we heard about it here in Amsterdam. It was great fun when the trophy was presented. The engineer took it well. He was delighted with his trophy, and he even placed it in a special place in his house when he returned to Wichita.

Final Modifications

The aircraft arrived in Amsterdam at the end of the summer. It was painted in Air France livery and had a French tail number. After they parked it in hangar four, our modification work began. I always felt sad when a brand new aircraft arrived, freshly painted, clean, and fully operational with just a few flight hours, and I had to open it and partly disassemble it. And that happened with 25266. Soon, we had the new tail number: PH-MCN. It took us six weeks to install all the wiring and reconfigure the aircraft. The Boeing team was already back in Wichita on a new assignment, and the Dutch team was modifying the aircraft. By this time, more than 100 people were involved in the process. Hats off to the Boeing engineers: They did a terrific job. There was not one mistake in the wiring diagrams, and they kept a very accurate wire list and hookup list. They also kept a connector pin configuration log, and all the details were perfect. This was very important because it was the reference material. Our ELMO (Electrical Modification) team might make a mistake, but they would at least have a good wiring diagram as a reference to troubleshoot the system.

In the last week of the project, one of the Boeing guys came back to help us with troubleshooting. He looked like John Denver, and therefore, in Amsterdam he was known as John Denver. The final week of troubleshooting was rough because we were all working day and night. John Denver was not only an excellent engineer but also a superb troubleshooter and a great guy. After the ground test and flight test, the aircraft now known as MCN was released for service. Martin, the boss of Martinair, sent all the engineers and techs a thank-you letter and a bottle of wine.

The first commercial flight of the aircraft was from Amsterdam to Taipei. After that, Aircraft 25266 was in service at Martinair for the next 25 years (Figure 5.1). It began its career with the tail number F-GCBN in Air France livery. In 1991, it became the PH-MCN, and it was in service with Martinair until 2008. After that, it was sold to Midex and used the tail number A6-MDG. The last days of its life were under White-Walker Holdings LLC with the tail number N54VP. The airplane was stored after its useful life of 30 years. I was gratified to know that at least 30 of the modifications designed by us were used in the aircraft throughout those years. These were not SBs but my own designed drawings under the STC. I was immensely proud of that work. Best of all was that we had a dream team on the project!

FIGURE 5.1   S/N 25266 flying for Martinair for more than 20 years and then for Midex at the end of the lifecycle.

Courtesy of Piet Alberts.

Courtesy of C L Wong.

6

RVSM: Really Very Smart Move

The sky became very crowded in the early 1990s, and everybody was seeking a solution to the congestion problem. The industry was looking at the big picture and promoting the concept of "free flight." Popularly speaking, free flight is when you fly from point A to B using the most direct route and not the predefined waypoints and corridors: in other words, the shortest route. This sounds very attractive, but at that time, the technology was not in place to allow this. Then FANS became the buzzword. FANS stands for Future Air Navigation System. This was the software in the Flight Management Computer (FMC) capable of flying from point A to B without waypoints, just the shortest route like a free flight. Honeywell and Boeing designed the software, which could be purchased at an astronomical price. The cost was negligible compared to fuel savings when flying FANS routes. Our 747-400 test pilot said: "If we must fly the FANS route, we will have a business case to make the investment." And he was correct. Unfortunately, there was no FANS route for us. There were some FANS routes defined in the Pacific, but only Qantas and Air New Zealand were flying in that area.

Later, they changed FANS to CNS/ATM, hoping that people would be attracted to the name, which stands for Communication Navigation Surveillance/Air Traffic Monitoring. It was a complicated concept, and I started to call it Cool New Stuff/All it Takes is Money. It was difficult to explain to management in just one sentence and, therefore, difficult to sell. Then many people started to promote the Global Positioning System (GPS). GPS was supposed to be the new solution to congestion. The problem was that the system architecture was not settled. Some people were promoting a stand-alone GPS receiver, and others favored an MMR solution. MMR is a Multi-Mode Receiver, the LRU with slots for the Instrument Landing System (ILS), marker, VHF Omnidirectional Range (VOR), and GPS. In this situation, the ILS receiver (or VHF NAV receiver) was removed, and MMR installed. This was not a simple process, and the costs were very high.

In those early discussions, it became clear that there was one simple and cost-effective solution that was almost ready for use. That solution was RVSM, or Reduced Vertical Separation Minimum. The idea was to reduce the vertical separation above the North Atlantic for flights above 29,000 ft. The new separation would be 1,000 ft instead of 2,000 ft. When implemented, RVSM would make it possible for air traffic to be doubled above the North Atlantic. Unfortunately, there was one "but." There is a transition area on both sides of the Atlantic. You can pump up the traffic above the North Atlantic, but in doing so, you create congestion in the transition areas. RVSM just replaced one bottleneck with another. When the RVSM above the continent becomes a real possibility, there might be benefits in Europe and US/Canada, but the bottleneck will shift to Asia.

Seattle Meeting

RVSM started with some preliminary meetings, which were useful because they educated people, and, at the same time, they required coordination between the many parties involved. Because I was responsible for Air Data Systems for the whole fleet, I was the one who was supposed to run the show within my company, and I was required to attend all RVSM meetings. The first RVSM meeting was in Renton near Seattle. It was in the summer, and I was with my family on holiday in the US. The plan was to fly to Phoenix, rent a car and have a three-week holiday in Arizona, California, and Nevada, and then fly home from Phoenix. On our stop in San Diego, I left my family in a hotel in La Jolla, California, and flew to Seattle for a two-day RVSM meeting. The meeting was in one of the Boeing buildings in Renton. There 40 to 50 people attending, including representatives from a number of airlines, including Boeing and Douglas, and some OEMs. We had a lot of discussions because airlines didn't want to spend a lot of money, and some aircraft types were difficult to make compliant. Because we were all engineers, there was a lot of conversation about RVSM during coffee breaks, and it was easy to connect with people. At the end of a day of meetings, a small group of us (KLM, United, BA) decided to go to a restaurant for dinner. A United guy was gathering information about restaurants, and somebody told him that the best place would be McCormick and Schmick's at First Avenue, so we went there without a reservation.

The guy at the door told us immediately that the wait time was long, as much as 2 hours, and then he noticed the small pin on my tie. It was a golden Boeing 737 pin. He asked me: "Are you a pilot?" I thought, nobody knows me here. I could say that I am a Chinees acrobat. Who cares? So I said: "Yes, I am." Then he said: "Oh, sir, I am a private pilot. And I like flying. Wait a moment, I will get you guys a table. Pilots help pilots." And he got us a table immediately. After dinner, when we were leaving, we saw people by the door still waiting for a table. Sometimes luck is your best friend.

That meeting in Renton was the first of a series. It was just the beginning and my first encounter with RVSM. The next meeting was again in Seattle, but near the airport at DoubleTree Hotel. At that meeting, Boeing revealed the plan to do the skin waviness check around pitot tubes on the 747 aircraft. This was a sensitive topic, because the aircraft was fitted with pitot-static probes that had a static port at the probe itself. All other aircraft, such as

DC-10s or Airbuses, had a static port at the fuselage. (The port is not so sensitive to local airstream around the fuselage). Boeing was helpful and provided a good plan: The airline/operator would do the skin waviness measurement around the probes using a special tool designed by Boeing. In total, there were 144 measurement points per aircraft side, all of which had to be sent to Boeing. Boeing would run the Computational Fluid Dynamics (CFD) analyses and provide the OK statement. If the model failed, skin shimming was required, following Boeing's provided instructions. Of course, that service came at an extra cost. Some engineers complained nonstop about the process, spoiling the conference for other attendees.

Copenhagen Conference

The next major step was a conference in Copenhagen six months later. The RVSM project had matured by that time, making this perhaps the best RVSM conference I attended. All the important parties were there: the International Civil Aviation Organization (ICAO), Aeronautical Radio Incorporated (ARINC), FAA, Joint Aviation Authorities (JAA), National Air Traffic Services (NATS), Boeing, Airbus, Douglas, and nearly all the European airlines. If you wanted to be successful with RVSM, you had to be there. By that time, all the details were known about RVSM, and it was much more complicated than I had thought. It was an art to synchronize all the attending parties, and I began to realize how important ARINC was and how important coordination within the aviation community was.

All parties really started to work together. At that time, though we didn't know it, we had almost 2 years to go before RVSM was operational. I always said that the secret to finishing the project on time is to start on time. The importance of this truism is generally underestimated, and people tend to waste time at the beginning of a project, which forces them to run at the end to meet the deadline. What follows is the scope of the RVSM project.

FIGURE 6.1   RVSM Button we got from FAA in Copenhagen.

Courtesy of Marijan Jozic.

RVSM Project Plan

FAA published the guidance materials for RVSM, stating the rules and regulations. ATCs have to train ATC controllers for RVSM operation and especially for operation in transition areas. Airframers will provide the SB per aircraft type with instructions for operators on how to qualify aircraft. Finally, about 25% of the fleet must be measured using Height Monitoring Unit (HMU), which should be installed in the aircraft. The unit consists of a GPS receiver and a laptop. When installed in the cockpit, it measures altitude during the flight. Data are then sent to ARINC, and ARINC compares them with the ATC data of the same flight. If data are within the tolerance, the aircraft is good to go. To avoid all the fiddling with the HMU, the aircraft can also fly above ground-based Height Monitoring Radar. There are two such stations (Scotland and Ireland), which check the aircraft's altitude automatically. Unfortunately, not many aircraft had a chance to fly above those radars, so an HMU was a necessity. When 25% of the fleet is checked, it is declared RVSM capable. At this stage, pilots have to be made aware of the new situation, so a pilot briefing is required. Finally, I determined to have a small sticker placed in the cockpit after the aircraft is checked that says: RVSM approved.

St. Petersburg Conference

The only difficult aircraft type was the 747. Those birds had to undergo the skin waviness check, and some of them, perhaps 20%, would need the skin shimming. With such a plan, we went to the last conference, which was in St. Petersburg, Russia six months later. This conference was a bit messy. The biggest problem for attendees was that we had Russian-speaking presenters. This required the use of simultaneous translators so that we could listen to the translation in English through our headsets. The translators were not used to translating aviation technical English and were obviously not familiar with aerospace terminology and acronyms. Every now and then, the simultaneous translator would say: "Oh gee, I don't know the English expression." One interesting thing I remember from this conference was that the Russian Aerospace Agency Director was not called Agency Director but "First Navigator." Someone introduced him like this: "Ladies and gentlemen, our keynote speaker is the First Navigator of Russian Aerospace Authorities." It seemed kind of strange to call somebody First Navigator—like we were on Star Trek or some other science fiction show.

St. Petersburg is a lovely city full of history. A group of us went for a walk in the evening hoping to find a good restaurant. On one street we found a restaurant with a Russian playing a trumpet at the door. He asked us where we were from, and we said Amsterdam. He immediately started to play the song "Tulips from Amsterdam." That locked us in on this restaurant. The food was good, but the price was astronomical. We paid about 600 USD on a meal for three people. The next day, we talked to the Russians at the conference about our experience, and they told us that we had eaten at the most expensive restaurant in St. Petersburg. Why me?

Qualifying Aircraft

Eventually, the process of qualifying aircraft started, and it went quite smoothly. But on one particular day, bad news arrived at my office: The pitot probe manufacturer had obviously underestimated the market. I had ordered new nickel-plated pitot probes for 747 aircraft and am now told they can't deliver them. That was my struggle until the end of the RVSM project: a battle with the OEM for each 747 probe. At least the fight started quite early, so we had time on our side.

My engineering order for skin waviness checks was progressing. Boeing was doing CFDs, and now and then, we had to do the skin shimming, but so far only on aircraft during a Heavy Maintenance Visit of the aircraft D check, so there was no impact on the operation. I also decided to paint the corners of the area around the pitot probes on the 747 red to indicate the sensitive area. No dents or skin damage is allowed in that area.

At this point, I had to go to Annapolis to ARINC to get one of those HMUs and begin checking aircraft. In Annapolis, I met a young guy who trained me on how to install the HMU in the cockpit and how to transfer the data via modem to their server. The next day, I went to JFK Airport to take on the first KLM 747 to be measured. The station manager got me into the aircraft before the crew and passengers. I had to install the GPS antennas on each side of the cockpit. The antennas were on a plate that had suction cups. Both antennas were connected to a laptop, and the laptop was connected to a socket just below the flight engineer's panel. The HMU was activated, and until we reached Amsterdam, I didn't have to do anything else with it. The crew was aware that I would fly with them in the cockpit, and everything went smoothly.

Once in Amsterdam, I cleaned up my stuff and went to my office to send the data to ARINC: simple and easy. A couple of weeks later I got an angry message from the administrative assistant of our chief executive officer (CEO). She was complaining that my telephone bill was too high—the highest of all 5,000 Engineering and Maintenance employees. I had to explain RVSM to an admin who didn't believe me: This is the cheap Dutch mentality.

End of the Story

The deadline was approaching, and only one aircraft was not compliant. It was the 747 with a tail number BUK, which can still be seen in the Dutch Aerospace Museum Aviodrome. The OEM was not able to deliver four pitot probes on time (Figure 6.2). They were on the way to Amsterdam but did not arrive. I arranged with the aircraft planning department not to schedule the aircraft for a flight to Chicago but rather to Kansai International Airport (KIX) (Osaka) and informed our flight operations (flightops). No problem: The flight to KIX is not within the RVSM area, and the 747 can fly. By the time that aircraft comes back, the probes will be in our hands, and we can replace them before the next flight. After that, there will be no restrictions and everybody will be happy: I can declare the project finished.

FIGURE 6.2 747 short strut pitot static probe looks simple from outside but it has very complex technology inside.

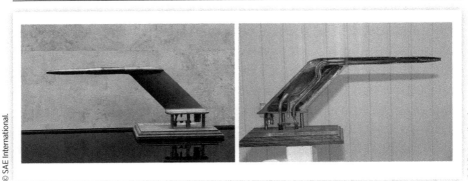

Now for another "but." The head of engineering was informed in a meeting about that flight to KIX, and he started to panic. He immediately started to shout that the RVSM project was one big failure. Managers were coming to my desk asking for an explanation. What could I say? The whole fleet was flying, all 150 aircraft of different types (MD-11, 767, 747-400, and 747 classic) were in the air or at airports worldwide. How could the RVSM project be considered a failure? After a discussion, the situation calmed down. The next day, the four nickel-plated probes arrived, and the day after, they were installed in aircraft tail number PH-BUK. The fleet was RVSM compliant.

Unfortunately, the last chapter of this story was not yet written. Two months later, I faced a problem with one 747. This 747 had just left on a flight after its yearly inspection. They got a warning from ATC that their altitude was incorrect. Upon return, the ground crew replaced the pitot-static probes, did the pitot-static leak test, and released the aircraft. On the next flight, they got the ATC warning again. The maintenance support center called me and asked for advice. After I heard the story, I suggested the skin waviness check and CFD. The next day, Boeing informed us that the aircraft was out of limits and that it needed the skin shimming. Well, a man's got to do what a man's got to do: Without skin shimming, there will be no flying. The whole process took us almost a week. The aircraft was released, and the altitude measurement was OK again. The only explanation was that the aircraft was damaged in the pitot probe area during the C check or immediately afterward. Otherwise, how do we explain that the aircraft was OK before the C check and problematic afterward?

Again, there is one more "but": The guy in charge of C checks suggested that management do preventive shimming during each C check every time. That idea was so ridiculous that I have no words to describe it. He had no idea what he was talking about. It was expensive to keep a 747 one week on the ground for skin shimming, but doing skin shimming on each 747 every 18 months (C check cycles) is crazy. I had to defend myself and the process to my managers, who easily got into panic mode.

The situation calmed down, and RVSM became business as usual. Nobody talked about it anymore; it was invisible. All of us who introduced it back in the 1990s were assigned to other projects, and many new engineers have never even heard of it. I am happy that I was among those RVSM pioneers. At least there is some information on Wikipedia about it. It is part of aviation history. And that expensive meal in St. Petersburg is now just a delicious memory.

7

SATCOM: Hello, Can You Hear Me?

There is no business like the avionics business. As with fashion, there are industry trends. There are trendsetters and trend followers. Back in the 1970s, the airline business was boring: not many developments, not many trends, and little excitement. After the introduction of microcomputers and microprocessors, a whole new technological world was discovered.

The trendsetters in the airline industry were the ones with guts, and the trend followers were those saying: I don't want to be the first guy to put that stuff in the aircraft. In the early days, everyone was pioneering, but some were more ambitious than others. The retrofit market was not yet discovered. Boeing, Airbus, and Douglas were only interested in selling new aircraft. Vendors were interested in selling the LRUs. Things got interesting after the introduction of TCAS.

After TCAS there was a gap. What to do next? Some people thought that SATCOM would be the next big thing, followed by GPS. There were airlines that dramatically miscalculated SATCOM. Their business case assumed that they would have 45 minutes of telephone traffic per aircraft tail number per day. Many years later, they were measuring only 10 minutes of telephone traffic per day. The system is very complicated and can be considered one of the greatest achievements in telecommunication engineering (Figure 7.1). But due to the exorbitant expense, SATCOM was and is almost never used.

FIGURE 7.1   With three such geostationary satellites there is full SATCOM coverage of Earth (except the north and south poles).

FoxPictures/Shutterstock.com.

What happened? Well, back in 1993 most TCAS modifications were complete. Suppliers knew that there were big bucks in modifications and installations of new systems. It was required for TCAS because there was an FAA mandate, and everybody was working together to meet that mandate. After that there was a gap to be bridged. Some suppliers immediately started to promote SATCOM. Airlines were looking to each other and waiting to see what would happen next. At one point, Lufthansa decided to install SATCOM on one of their aircraft just for a trial. Many European and US operators were monitoring the situation, and the big news arrived. You can make telephone calls from the aircraft and the flying public is actually enthusiastically doing that. They were using SATCOM and were very positive about it. Lufthansa said that the trials were free for passengers, but nobody paid attention to that part of the statement. Many airline operators were saying: Yes, yes, people like it; they want it! SATCOM would be a great success.

Immediately, the other airlines said: We must install it; otherwise, we will be way behind Lufthansa. But honestly, if it is free, every passenger will try it. If they have to pay, it will be a different story. This was logical, but nobody wanted to hear that argument.

At my company, they immediately hired a student to work on a SATCOM preliminary investigation project and tried to define the justification for installations. At that time, the costs of installation were unknown. People were talking about one million USD per aircraft, but nobody really knew.

The result of the investigation by that student was known before the investigation started. In short: Yes! We need SATCOM to prevent being left behind by other operators. Many people will use it. Tourists especially will use it because they will find it cool to call relatives from the airplane and say: "Mother, I am calling from the aircraft, and we are above the Atlantic Ocean." How cool is that!

The SATCOM Project

After some time, and after management got the student's report about SATCOM, it was decided to launch the project. Two of my fellow engineers got the project. Around that time, my airline and Northwest Airlines became partners. This was big news in the aviation world. There was a need for more close cooperation, and SATCOM was one such project, because Northwest was also very interested in SATCOM. My fellow engineers had to go to Minneapolis for two months to work on a joint SATCOM specification and vendor selection. Only now, many years later, can I recognize that the project was a lot of baloney. The first big decision was supposed to be made between a top-mounted and side-mounted antenna. The second big decision was vendor selection. It was a prestigious project, so two months in Minneapolis for two engineers was not an issue. They would just go there for two months and discuss the project. After two months they decided that a top-mounted antenna would be the best choice and that a six-channel Honeywell system would be the best vendor choice. What a surprise.

In the meantime, I was at home helping Saudis install a three-channel SATCOM in their 747, which was in our hangar for heavy maintenance and SATCOM installation. While my fellow engineers were working on the business case, I had already collected the experience with SATCOM installation. This turned out to be lucky for me.

The SATCOM business case was eventually approved, and when the project started to get difficult, those two fellows were moved to another job, and I inherited their SATCOM project. This was the second time I had missed all the fun from the beginning of the project and was assigned the project when it became tough.

One of my first business trips after I got the SATCOM system project was to London to attend the Inmarsat conference. The takeaway was the statement of an Inmarsat hotshot who told the crowd that it was ridiculous that he was on a flight between Singapore and London and couldn't make a telephone call. Therefore, every self-respecting airline should have SATCOM and a telephone on board. He was obviously trying to sell the idea of a telephone connection on board. Even 30 years later, people are not making telephone calls from the aircraft. Have you ever seen somebody calling during a flight? I collected a lot of flight hours on wide-body fleet, but I never saw anyone make a telephone call.

The SATCOM design, installation, and certification was a great challenge. It was expensive, new, complicated, and risky. Not only the modification but also the support after installation of the system was a great challenge. Only a few systems in the entire world were flying using the brand-new design, so no one was able to share their experience.

After every installation, we had to do the commissioning of the system. This was new in aviation. Usually, we installed a new system, completed the test, and signed off on the paperwork. But Inmarsat asked for a special procedure to verify that the system was working. Commissioning is nothing more than parking the aircraft outside after the installation, after which you call the ground station and tell them that you want to commission the aircraft. Most of the time this was done in the evening. My installations were combined with C checks. C checks were finished on Friday evening, and the aircraft was pulled from the hangar to a suitable spot to allow it to "see" the satellite.

The operator of the ground station was usually asked to do the telephone calls on all six channels. Then they do the measurements of the signal and sign off the commissioning. On one occasion, I decided to use one of the channels to call my father in Croatia. It was already 10 o'clock in the evening in Croatia. I got him on the line and enthusiastically tried to explain: "I am calling from the 747-300 aircraft and the signal is going to a geostationary satellite and back to GES (Ground Earth Station) in France. After that, the call is connected to a terrestrial network and eventually to your phone in Croatia." He responded: "Can't you leave me in peace? It is 10 o'clock, and I am already in bed. Call me tomorrow." That was the end of my first SATCOM call.

Troubleshooting

Troubleshooting experience was essential to my work, and a whole new way of testing and troubleshooting was introduced during this project. The laptop became a powerful engineering tool. Without a laptop, it is impossible to troubleshoot SATCOM. Once the engineer is used to using the laptop, the procedure is simple: Just connect the laptop to the Satellite Data Unit (SDU) and start the "ProCom+" (less-experienced techs can also use the CMT program [Computer Maintenance Terminal] delivered by the vendor, which is more user friendly). Hundreds of fault codes could be pulled from the SATCOM system and the fault could be isolated. The BITE (built-in test) was magnificent. Honeywell engineers had done a good job. During the design of the system, all possible faults were analyzed, and software was developed to help engineers in their troubleshooting task.

For the airline engineer, this was a step into a whole new technology. Suddenly, they don't need a voltmeter or a spectrum analyzer but just a laptop. They had to start thinking differently. With computers, it's all about the software. And there are some new rules as well: The engineer must learn to interpret the fault codes. Those codes are very important and will guide the engineer to the failure. Using their laptop, they can ask SATCOM SDU for the current failures. The SDU will respond instantly. If the engineer is not experienced, the long list of failures reported can be overwhelming.

Even with a long list, it's not that bad. In no time, the engineer can start to interpret the codes to find that some of the issues can be eliminated or masked. SDU will, for example, report that the system is down because the INS input is not present. This can mean that INS (Inertial Navigation System) is not lined up. After the INS is lined up, the code is cleared and the engineer can concentrate on the next fault code. After some time and experience, the engineer can start to play with the system. OK, let's check how many telephone calls have taken place during the last six weeks. When was the last one? At which position was the airplane at that time? What was the elevation of the satellite? And so on.

What a great time it was! If a SATCOM failure becomes difficult to explain and correct, no problem. Just download all the data from the system (for experts in Honeywell SATCOM system, this is known as "log all"), bring it to your office, and email it to your counterpart engineer at Honeywell. He or she will almost instantly come back with some good ideas on how to tackle the problem.

In that high-tech environment, everything seems under control. But now and then, there are some extreme issues. That is when the engineer's job starts to become interesting.

I should probably explain the SATCOM system architecture. The SATCOM rack at the 747 aircraft is known as the E-42 rack. It is located above the ceiling approximately halfway through the aircraft. SDU, RFU (Radio Frequency Unit), HPA (High Power Amplifier), and BSU (Beam Steering Unit) are installed side by side on a strong aluminum E-42 rack.

The modifications on the entire 747 fleet (classic and -400) were finished. After six months of use, one by one, the aircraft were having issues. Pilots were complaining that immediately after the aircraft was airborne, the SATCOM system was announcing "log off," meaning "out of business." This was obviously an intermittent failure. If they cycled the CB (Circuit Braker), the system would "log on," meaning back in business, and after some time, it failed (logged off) again. The same failure was later announced on at least ten aircraft. Connecting the laptop, the historical failure log showed "loss of min ops" and provided a list of faults related to the antenna system. On every aircraft a different failure pattern was recorded. Sending the data dump to the vendor and asking for advice was useless.

Now the search for the problem began. Line maintenance people were replacing BSUs, SDUs, and HGAs (High Gain Antennas). The vendor of the BSU and HGA provided the shop findings and complained: The antenna was always NFF (No Failure Found) and the BSU input transformer was almost always burned.

The poor techs and I, as the engineer, were in a difficult position. Our high-tech tools were not much help. At one point, one of the 747s was grounded, and I received the following message: Airplane is AOG (Aircraft On Ground). Make it work, and make it work fast. The airplane is not going to fly until the SATCOM is repaired. Two years ago, SATCOM was not yet installed, and now it is a no-go item. That was the new world.

The SATCOM always worked on the ground (except when the BSU was burned and not operating). After 10 hours of trouble-shooting, there was still no good news. The system was checked and re-checked, switched on and off, and tested and re-tested without any significant success.

Unfortunately, in the high-tech SATCOM environment, I didn't think about the first law of engineering: "When something doesn't work, kick it." I should have done this first, before spending 10 hours trying out high-tech tricks with the laptop.

Well, after a hard kick against the E-42 rack, the LED (Light Emitting Diodes) of the BSU flickered. That was my Eureka

FIGURE 7.2 There is a lot of wiring in the aircraft and wire chafing might be a big problem.

moment. It's the wiring! I was convinced that it was the wiring, but the actual problem was difficult to find. It turned out that the cable bundle from the BSU to the terminal strip was chafing against the aluminum beam. On every 747 a different wire was intermittently shorted to the ground. On the wire, the section with chafed insulation was almost invisible. It was just a tiny, little gray spot that looked like a dust particle. An inexperienced person would have overlooked it for sure (Figure 7.2).

The fix was easy: A Teflon sleeve around the bundle and an insulation strip at the aluminum beam. Costs of material: 3.00 USD. During the next C check, the bundle will be made shorter, and the damaged wire will be replaced. This serious fleet problem was solved: No more antenna removals, no more burned transformers, no more failures.

Having lots of experience and knowledge is worth little if you don't share what you've learned with others. And lessons learned have no meaning without corrective action. The real corrective action was not the installation of the Teflon sleeve or repair of the cable bundle. The real corrective action was another issue: the education of people working on the airplane.

The technician who installed the wire bundle was not aware of possible chafing due to vibrations. Even if the cable bundle was at a safe distance from the aluminum bar on the ground, it is a different story in the air. (Note: Chapter 20 of AMM [Aircraft Maintenance Manual] is dealing with standard practices and there is a lot of information about repairs of wiring). Even if the engineer specifies a wire kit with longer wires, the tech installing it must know that there is a danger of chafing. A too-short wire is not good; a too-long wire is even worse. It's a matter of good workmanship. Even the inspector, when checking the work, didn't notice the problem. It comes down to awareness. If the tech and the inspector were aware that vibrations can cause tricky problems in the aircraft, they would know to make the cable shorter, but not too short. Awareness can be learned in school or the hard way. The hard way is always more expensive. The SATCOM failure described above will not happen again.

Even in a high-tech environment, it all comes down to the avionics people. The above-described SATCOM failure cost the company at least 400K USD (unnecessary removals of antenna, repairs of BSU, no failure found on SDUs, 1-day AOG [think about the loss of revenue for one day of a 747 operation], work hours for corrective action, meetings, telephone conferences, unnecessary purchase of extra spare antenna and BSU, etc.). Ask yourself how many people might be sent to a course to learn some tricks about wiring and wire installation on that 400K USD (For a better understanding see AMM Chapter 20).

Mr. SATCOM

After this, I became Mr. SATCOM. I had missed the fun in Minneapolis with the Northwest engineers in the specification phase of the project, but I became possibly the most experienced SATCOM engineer at the airline. This was evident when I got a call from the Dutch Air Force. They had a government business jet equipped with SATCOM (Gulfstream IV), and there was a problem: They were able to receive a fax in the aircraft but were unable to send one. The Dutch prime minister had been in Egypt, and they had an engine failure. The aircraft was grounded, and the prime minister had to return to The Netherlands on a commercial flight. The crew and repair techs were now having a problem sending documents via aircraft fax. They were upset that communication in such a situation was unreliable. One of the generals decided that the airplane was not going to fly until the problem was resolved.

The airplane was grounded at the Air Force base, and now the maintenance crew had a problem: They didn't know what to do because the system was new, and no one was trained to work on it. They called Honeywell in Phoenix. Honeywell told them that "there is an engineer with the name Mr. Jozic who can probably solve the problem. Since he is not far away from the airplane location, that's the easiest way to get an expert on the airplane." This is how I got involved. That same day, 3 hours later, I arrived with my laptop at the Air Force base. They immediately brought me to the airplane, and I started the standard troubleshooting procedure using my laptop. After some time, it became obvious that the SATCOM was OK, so the fax was the next suspected device. It was 2 o'clock in the morning when the failure was located. The fax machine had a built-in delay. If a dial tone was not detected within 5 seconds, it would time out and terminate the connection. That delay was adjusted to 15 seconds, and everything was A-OK. We could send faxes in both directions and make telephone calls. Everybody was happy. After a long day of successful troubleshooting and repair, we took the opportunity to empty the aircraft refrigerator and drink all the VIP beer. Oh how I love the Dutch Air Force (and VIP beer!).

Final SATCOM Thoughts

SATCOM is a great system: Always available, always perfect. Unfortunately, the idea that passengers would use it more often was simply wrong. Inmarsat was playing games and wanted to earn big bucks, but it didn't go well for them. They lost their momentum due to the steep cost of SATCOM. The first minute was 15 USD and every additional minute was 10 USD. Rather than spend so much money, most people preferred to wait for the aircraft to land to make a call. By the time Inmarsat realized how wrong they were about SATCOM, mobile phones were in use everywhere. People started to send text messages instead of calling. You could also see passengers waiting for their luggage and making telephone calls. For a less-than friendly price, passengers could call relatives before landing. It was much smarter to wait and use an inexpensive terrestrial network than spend money on expensive calls from the aircraft during a flight. Making 45 minutes of telephone calls per tail number per day was never achieved. I could see this on my laptop when I download the maintenance log from the SDU. The average was 10 minutes per tail number per day, and this was mostly from the cockpit. Even the crew was not using SATCOM very much. Fearing overuse, some airlines didn't want to open the system to the cockpit crew. Often, only the company prepro-grammed numbers could be dialed by the airline crew. Management was afraid that the captain would call their girlfriend/boyfriend on the company dime. But even that was not happening. I had a lot of fun as the engineer responsible for the SATCOM system, and that is what matters most to me.

8

MLS: The Last Days of the Microwave Landing System

I was not there when they designed the ILS, MLS, and GPS. I was also not there when they issued the ARINC specifications for MLS (ARINC 727). But I was present to help write the final chapter of MLS.

There were two MLS-minded people in Europe, and when this chapter begins, they are nearly ready to retire from their airlines. Before they retired, they wanted to introduce the MLS in their company. The next generations would remember them as the people who introduced MLS as a new system for landing in bad weather. For these individuals, it was not an issue if 2 million (M) USD should be spent per aircraft with a minimum number of 200 aircraft—in addition to 1M USD for airport installation at each runway. They didn't care about the cost as long as they got their toy: MLS. Luckily, the companies were unable to afford that toy.

ILS

This story begins in the first half of the last century with the development of ILS. At the time, aircraft approaches and landings in poor weather conditions were risky (Figure 8.1). Early air navigation systems provided only basic lateral position information. In the 1930s, airline pilots flying in low-visibility conditions had to take a heading off a radio navigation station and then use speed, time, and distance calculations to figure out when they should see the runway at their destination. Today, most pilots would not even

FIGURE 8.1   Landing in bad weather.

think about doing that. In the old days, descents were dependent on the accuracy of a pilot's calculations. All final approaches had to be visual because no instrument guidance system could direct a pilot on a safe, precise descent.

The successful demonstration of a full ILS in 1937 marked a great advancement in air transport operations. For the first time, pilots could tune in to a radio signal transmitted from an airport runway and receive precise lateral and vertical guidance on an approach and landing. The ILS broadcast a straight, narrow VHF/UHF signal beam that started at the runway and rose at a three-degree angle. By centering the vertical and horizontal needles on the ILS indicator, a pilot could fly down the ILS "beam" to the runway, descending on a three-degree glide slope that was lined up with the runway centerline. This system improved the safety of airplane travel and improved operations in bad weather conditions. US airlines soon began using ILS approaches on a regular basis.

By 1949, ILS had become the world standard for landing systems. However, air traffic congestion and the need for more noise-free approach paths to airports created demand for a more capable and flexible landing system. One problem with ILS is the limited number of channels—40. As air traffic increased, planners began to realize that certain areas of the country would have more airports and request more ILS frequencies than the ILS frequency range could accommodate. Also, the relatively low-frequency range of ILS was sensitive to signal reflection, or "multi-path" errors. Another limitation of ILS was that it allowed only one approach path to a runway. This is a very important issue. The receiver in the airplane would lock at the ILS "localizer" and "glide slope" up to 10 miles away from the runway and fly a straight course in for a landing.

In 1971, two of the top priorities identified by the FAA and NASA were to:

- Increase the air traffic capacity of airports.

- Develop approaches that avoid noise-sensitive areas to reduce noise pollution. NASA first suggested curved approach paths to enable the FAA to accomplish both goals. But this approach would require a more flexible landing guidance system than ILS. This was supposed to herald the end of ILS, but even many decades later, no one even dares to consider decommissioning ILS systems.

Basically, there were two ways of designing a scanning beam microwave system:

- "Doppler Scan" technique, using modulations in frequency to tell the pilot the position of the airplane in relation to the runway.

- "Time Reference Scanning Beam" (TRSB) technique, known today as the Microwave Landing System (MLS), which uses time-referenced sweeps of a single frequency to pinpoint an aircraft location.

Doppler vs. TRSB

Throughout most of the 1970s, major national and international organizations, including the FAA, NASA, British Civil Aviation

Authority (CAA), ICAO, and Lincoln Laboratory, struggled to develop a standard MLS from these two competing technologies. In April 1978, the All-Weather Operations Division of ICAO voted in favor of implementing TRSB MLS. This decision was politically charged. There was a lot of money involved and because of that there were numerous discussions about who would pay for development and who would take the lead on the project. The conclusion of this process was that the best possible replacement for ILS is MLS because of its curved approach capability, which can increase capacity and lower noise pollution. The plan was to decommission ILS and install MLS instead.

After their April 1978 decision, ICAO began to develop Standards and Recommended Practices (SARPS) for MLS. The target date for switching over to MLS as a primary landing guidance system was then changed from 1985 to 1995. The discussions about financing were still on the agenda and there was no hard evidence that ILS was degrading. Though delayed, the future of the TRSB MLS seemed secure. The technology was proven, and ICAO had given its stamp of approval. After the SARPS were written, all that remained was to produce and install the equipment. At this time, ARINC Characteristic 727: Airborne Microwave Landing System standard was adopted by the Airlines Electronic Engineering Committee (AEEC). The latest version was published on August 27, 1987. Here is a quote from the document:

Equipment manufacturers should note that this document aims to encourage them to produce maintenance-free, high-performance equipment rather than that of minimum weight and dimensions.

The MLS path started to shift in an unwanted direction, and progress slowed considerably. Fifteen years later, only a handful of MLS installations had been completed. It was uncertain whether the MLS would ever be fully implemented, at least in the US. MLS encountered so many delays and obstacles in the 10 years following the ICAO decision that it was eventually overshadowed by newer technology, the GPS. GPS became so popular that a lot of people (most of them not properly informed) started to predict that the development of a GPS landing system was just a few years away.

In the latter half of the 1980s, people involved in aviation started to communicate more intensively about a possible better and cost-effective solution coming with the development of GPS landing system (GLS). It seems that every opinion was considered. Those in control of the money were confused because one group was pushing for MLS and another group was saying: Wait and see! The third group was convinced that GPS was the best solution. Without consensus, the money would not flow. No one wanted to bet on the wrong horse and have to spend money twice to fix the problem.

MLS was described as a temporary step from ILS to GPS. This was a fatal blow against MLS: The idea that MLS was a bridge between the old ILS and the brand-new development of GPS made MLS redundant before it was implemented. The conclusion of most airline executives was, "Why should I invest a huge amount of cash in MLS if it is a temporary system? I don't want to be the first to pay for that." ILS was dying, and now it seemed that MLS had started to die, as well.

The FAA awarded the first production contract for MLS to Hazeltine Corporation in Commack, New York, in January 1984. The specifications for the equipment required only a Category I level of approach performance. Category I conditions were the least severe of the three low-visibility condition categories, requiring at least 200 ft of cloud ceilings and a half mile of visibility. The guidance signal of the MLS equipment did not change from Category I to Category III, but Categories II and III installations required a much higher level of redundancy in the equipment. That was a fatal stab in the back for MLS. No major airline was flying to those airports with only Category I capability, and the critical mass of users would never be reached. The FAA planners hoped that by installing the first MLS at small airports, they would begin to build a consensus of support for MLS among regional airlines and corporate aircraft operators. This was a mistake on the FAA's part. Small operators didn't have the money to implement MLS, and big operators didn't need to fly to small airfields.

The problem with the FAA's strategy was that installing Category I equipment at outlying airports offered no opportunity to show the major airlines any of the more impressive advantages MLS could provide, such as more efficient approach procedures or better automatic landing capabilities. More important, MLS technology was no longer the only alternative to ILS.

Some of the problems with ILS had been rectified, and the growing financial concerns of the airlines following deregulation made them reluctant to purchase new avionics for their entire fleets unless they could see immediate benefit from the new equipment. But they were seeing nothing about MLS. Many US airports didn't have the same problem with mist as European airports, and that was one more blow to MLS. ILS was still OK, and the airlines could wait for GPS to mature.

GPS

In 1983, the US decided to make its satellite navigation network, called the Global Positioning System (GPS), available to the world. The system, which was developed originally by the Department of Defense for military purposes, was based on a constellation of geosynchronous satellites. By using satellite-transmitted radio signals to measure the distance from a receiver to several different satellites, the exact position of the receiver can always be determined. GPS offered so much more capability and flexibility than any other navigation system that it was often described in the literature as the greatest opportunity to enhance aviation system capacity and improve efficiency and safety since radios and radio-based navigation was introduced in aviation. GPS is not dependent on any ground navigation aids, which means that it can provide guidance in a much wider variety of locations than navigation systems based on radio transmissions from ground positions. GPS also provides accurate position and velocity information in three dimensions, at any instant in time, which other navigation systems could not do.

There were limitations to the GPS system, however. To keep the defense of the US from being compromised, the Department of Defense intentionally degraded the accuracy of the satellite signals for civil use. As a result, the position accuracy was only within 300 ft, which was acceptable for navigation, but not precise enough for a landing system. These position errors could be corrected, and the inherent accuracy of the satellite signals could be improved further with "differential" techniques, although this approach did require ground installations. In a differential system, a stationary GPS receiver was placed at a surveyed location near the airport. The ground site would receive the satellite transmissions, compute how much correction was needed to match its surveyed location, and transmit that correction factor to approaching airplanes through a real-time data link. One correction transmitter per airport would be sufficient. There were still questions about the capability of GPS to provide enough accuracy for the most demanding instrument approaches, and, of course, the integrity of GPS was questionable. If the constellation of satellites fails, there is no redundancy.

ICAO decided to wait until 1995 to determine whether to alter its plan to implement MLS. The original deadline called for installation at all international airports by 1998 and transition to MLS as the primary navigation system by 2000. By 1995, the FAA and NASA were expected to have substantially more information on the potential and limitations of GPS. ICAO members could then decide whether they wanted to extend the deadline for MLS or shift the main navigation and landing system focus to Global Navigation Satellite System (GNSS). The landing system decision remained a highly charged issue, however, and there were split opinions both within ICAO and within the US.

The UK decided to proceed with MLS implementation without waiting for additional research into GPS. Poor visibility conditions in the UK made the ability to perform Category III landings a necessity, and ILS frequency congestion and interference problems were more severe in Europe than they were in the US. The British reasoned that even if GPS technology could be developed far enough to provide the accuracy for Category III operations and the questions regarding system integrity, reliability, and availability could be resolved satisfactorily, the development and implementation of the system would take too long—a solution was needed quickly.

The US looked more favorably upon GPS or GNSS, but there were still strongly divided views within the FAA, NASA, and the aviation community about what the next landing guidance system should be. Most people agreed, however, that even if MLS was implemented, it would likely be in a much more limited capacity than originally envisioned.

The 25-year-long controversy over a new landing guidance system underscored just how much political, financial, and other external factors could influence the acceptance and application of new technology. In the mid-1970s, MLS clearly demonstrated its superiority over the existing ILS. Yet, almost 20 years later, ILS was still in use, and MLS was being overshadowed by more advanced technology. This is the history of landing systems. We are at the beginning of the 1990s, when the final chapter of MLS began.

Final Chapter of MLS

BA started to complain that there was a problem at Heathrow Airport. The ILS beam was not stable. Even British authorities CAA installed special shades/plates on the window of their building in the vicinity of the runway. These plates are designed to disperse the electromagnetic signal coming from the ILS transmitters (localizer and glide slope). The group of engineers at KLM flight ops heard about the Heathrow complaints and decided to review the situation at Schiphol-Amsterdam Airport. The conclusion was that, by 2004, Category III landings would no longer be possible. They looked to BA for decisive action, and BA looked to KLM. There was no news from Lufthansa or Air France.

At this time, there was a project called RVSM, which was meant to introduce greater airplane capacity above the North Atlantic by a change in vertical separation. Project TCAS was also running, and there were other projects with operational impact (EGPWS, GPS, etc.) as well. Nearly every large airline established a group of engineers who would follow these developments and provide advice to management. I was a member of that group at KLM. At 30 years of age, I was the youngest member and had no rights. It was an old boys club with a few youngsters who were there to dig into the books and do the boring work. Any talking was reserved for the old boys. Sometimes I thought that a few old engineers from KLM and BA decided to push for MLS just to ensure they were mentioned in airline history books.

For MLS, there was already ARINC 727 specification, but it was decided that this was not sufficient. A concept based on a dedicated MLS receiver was discarded. What, then, was the best replacement for ILS?

Many airlines were looking for a way to ameliorate the financial losses caused by the inability to operate in all weather conditions. Especially in Europe, there are quite a few ILS Category III operations. There were predictions that airports such as Heathrow-London and Schiphol-Amsterdam would suffer under ILS degradation. The best possible replacement for ILS was supposed to be MLS because of its curved approach capability, which would increase capacity and lower noise pollution. Predictions were that after the year 2000, Category III operations at Schiphol and Heathrow airports would not be possible. Airlines flying to those airports were lobbying for MLS. In 1995, the stand-alone MLS unit (built per the ARINC 727 standard) was judged to be unsuitable. A new design was necessary because, in most aircraft, there was not enough space to accommodate three more units in the Main Equipment Center.

The best possible solution was to integrate MLS and ILS in the same box, but some aircraft had ILS and VOR (VHF Omnidirectional Range) in separate boxes. Another concept was to have ILS and VOR in the same box. By that time (the mid-1990s), there were digital ILS and VOR boxes and analog LRUs. It was decided to specify a brand-new box called the MMR (Multi Mode Receiver). Unfortunately, it was not possible to build one box that could be fitted into any aircraft. It was necessary to design digital and analog MMRs. The digital MMR had, at that time, just an ILS card and empty slots for future growth to MLS and GPS.

The analog MMR was a replacement, for example, for the Collins model 51RV-5B VHF-NAV (VHF Navigation) receiver. Therefore, the MMR was designed to have ILS and VOR as the basic option, with empty slots for MLS and GPS. The flight ops departments of some European airlines started to lobby for a retrofit of perfectly working ILS receivers with MMR ILS only. In their opinion, it was a piece of cake to later add the MLS card in the MMR and then add a couple of wires in the aircraft, and that would complete the MLS installation. This was almost impossible to sell to management, because it was just a replacement of the existing functionality (ILS) with the same functionality in another box. And everybody knew that the MLS card had still not been designed.

Under pressure from some European airlines, it was decided to install the MLS ground station on two runways at Schiphol Airport. Because it was expensive equipment, the authorities decided to increase the landing fee to cover the cost. Airlines paid a higher landing fee for at least seven years due to MLS, and no one could take advantage of the system, because not a single airplane was using the technology. The idea was to encourage airlines to install equipment on board—of course, MLS installation cost 1M USD per runway. At the time, Boeing and Airbus were not offering an SB for MLS retrofit. Without Boeing and Airbus support, resources, and data, it was impossible to install MLS on the aircraft.

The management of BA and KLM mobilized to put Boeing under pressure to provide a price for MLS installation. Boeing did the estimate, which was about 20M USD. This was a lot of money, and therefore it was decided to establish a team that would define the architecture of the system and, after that, recalculate the costs.

There was also a meeting with McDonnell Douglas in Amsterdam. (This was before the merger of McDonnel Douglas and Boeing). In that meeting, I made the presentation. The takeaway was, "Schiphol airport is sick. The ILS is degrading. The cure for this sickness is MLS." That statement was nothing special in my mind, but 30 years later, I met a man, now a Boeing manager, who was in that meeting. He could still remember my story about the cure for ILS. I couldn't believe that when he told me 30 years later, but it was a contentious issue.

The ultimate goal of the group was to investigate the possibility and costs of MLS installation on Boeing 747-400, 757, 767, 737, and MD-11 aircraft. KLM and BA were, at that time, the only airlines willing to take the lead and retrofit part of their fleet with MLS. I was spending a lot of time in Seattle digging into documentation and having hundreds of coordination meetings. I learned a lot and count this as the period when I became a real, die-hard engineer. A great opportunity came my way, and I was not afraid to take it. Before those meetings I didn't realize that I was actually a very good engineer. My managers, who were coming to Seattle every three weeks for TCM (Technical Coordination Meetings), also noticed my knowledge and my way of discussing the subjects. We believed we were on the right track.

We had many discussions and brainstorming sessions to establish the definition of an installation. An Operational Concept Document was compiled to redefine and simplify the MLS concept. It was decided not to develop the curved approach concept. The easiest solution was to have MLS operate like the ILS straight approach. Some people call this concept ILS Look-Alike or ILS Similar. All our dreams about curved approaches were set aside. The next step was to eliminate Precision Distance Measuring Equipment (P-DME) for calculations of distance to the runway. Instead, the airport's inner and outer Marker Beacons would be used. A design goal was that the operation of ILS and MLS would be transparent to pilots, who would have a selector switch in the cockpit with two positions, ILS and MLS. In the future, the position for GPS could be added.

At the airports, the runways would be equipped with ILS and MLS, which would always be active. Pilots would make a choice and select the ILS or MLS type of landing by operating the switch and selecting the ILS frequency or MLS channel required for that runway. The computers would do the rest. The system would be designed in such a way for it to be impossible for one pilot to select MLS and the other pilot to select ILS. To make the system transparent to pilots, a lot of engineering work was required.

For example, for the 747-400 to be able to use MLS as its airborne landing system, the following actions would be necessary:

- All existing equipment should be modified to Boeing 1999 baseline; that is, all p/ns of avionics should be upgraded to the Boeing 1999 standard.
- Replace the ILS receiver with MMR (ILS/GPS/MLS).
- Modify Electronic Flight Instrument System (EFIS) displays to show MLS channel number if MLS is selected, or to show ILS frequency if ILS is selected.
- Modify the Flight Control Computer to accommodate antenna switching.
- Modify the Mode Control Panel to be able to select the type of approach (selector switch: ILS, MLS, GPS-based Landing System [GLS]).
- Modify the Central Maintenance Computer to be able to handle MLS fault codes.
- Install MLS antennas and cabling.
- Upgrade ADS (Automatic Dependent Surveillance) to MLS.
- Modify the FMC and Multifunction Control Display Unit (MCDU) to handle MLS data.
- Add appropriate wiring.

An extensive certification also would be required to demonstrate to the FAA and JAA that all systems worked as expected and that the airplane was capable of accomplishing Category III Autolands. It was estimated that at least 50 Autolands would be required for certification, which would necessitate at least two weeks of flight testing.

It was obvious that MLS implementation would be a gigantic project. Software in many boxes would be changed and certified. New p/ns would roll out. Integration tests would be done, and millions of small issues would have to be resolved before any flight. The Non-Recurring Engineering (NRE) cost estimate was so high that it was an unaffordable undertaking for just two airlines with approximately 200 aircraft. We are talking huge NRE costs. And honestly, if you invest an average of 2M USD per airplane and you use MLS just three times per year at Heathrow or Schiphol airports, then you have made a lousy business case.

How did we come up with three landings per year at Schiphol? A team was assembled to establish the Return of Investment (ROI) for implementing MLS. That team had a special breed of person. Their title was Decision Support Manager. This was their approach: They purchased a database of weather conditions at Schiphol Airport over the last 30 years. The weather was segmented into half-hour intervals. They developed software to calculate the reference year. The best fit was, I believe, the year 1996. Then they took the data from another database to find out how many passengers would miss their flights if the airport was a Category III condition (NOC = Not connected). For Category III conditions, the time separation is different: You can't land an airplane every minute. The time separation will move to 5 or even to 10 minutes. This means that the airport can't accept all planned traffic, and, due to delays, some passengers will miss their connecting flights. It's as simple as that. The result of this exercise showed that in such a reference year, the airline would suffer losses of 1M USD per day, and this would happen two to four days in such a year. The reader can very easily calculate the ROI. If you invest 200M USD to modify the fleet and save just 4M USD per year due to NOCs, it will take 50 years to recover the investment.

Based on this information, KLM and BA decided to pull the plug on the MLS project for their Boeing fleets. It was a wise decision.

And what about Airbus?

I was told that, in 1999, Airbus advertised that if a customer bought an A320, Airbus would equip it with MLS for only 20,000 USD. This story might be true, but in 2003, there was not a single Airbus airplane delivered with operational MLS on board. Remember that everyone was convinced that in 2003 the Category III landing would be impossible at Heathrow and Schiphol. Even if Airbus A320s were delivered with operational MLS, where could the operator use it? At Amsterdam's Schiphol Airport in 2003, of the two MLS ground stations, one was decommissioned and removed, and the other was switched off. The future of the Heathrow MLS ground station is less certain, because the British CAA still secretly believes in MLS.

What about the other airlines? No one is interested in MLS because no one feels any financial pain thanks to ILS, which is not degrading as was predicted. The development of MLS software and the design of MLS systems on Boeing aircraft would take another two, three, or more years to implement after the first airline decided to go for it. The airlines are not looking that far into the future. The only way to get MLS in use worldwide is to issue a mandate, decommission ILS, and force all airlines to comply with the mandate or go out of business. The FAA is not going to do that—they already stopped all activities regarding MLS. The JAA (EASA, European Union Aviation Safety Agency) is not going to do it either, because the member countries do not favor MLS (except for possibly the British CAA). The rest of the world followed suit.

One final thought: In 1995, someone did the calculations and investigated the quality of ILS at Amsterdam's Schiphol Airport. The conclusion was that after the year 2000, Category III Autolands would not be possible at any of the runways. As I write, we are now in the year 2023. At Schiphol Airport, Category III Autolands are still possible on all runways, and the hard-working ILS equipment continues to do its job. The air traffic controllers complain that, in misty weather, congestion occurs on the taxiways because if too many aircraft land under ILS Category III, they cannot see these aircraft from the tower through the mist on the taxiway in order to help them find the gate. The traffic jam has moved from the approach to the taxiway.

Many MLS lovers are retiring, or are deceased, and the new generation of engineers believe in GPS. I don't want to provide any predictions about the future of MLS, but I am afraid its last chapter was written at the end of 1999. It was close, but not close enough to get it done. I am happy that I was part of the process and that I can tell the story.

Many airlines are now in poor financial condition, and MLS is not on their shopping list. By the time MLS gains first place on the shopping list, GPS will be ready. We all know that at the end of the road (or the runway) is a GPS-based landing system. We also know that, at the moment, ILS Category III is still in good shape (at least at most airports). The airlines cannot afford MLS as an interim solution when they know that the permanent solution is GPS. It all started with the MLS target goals:

- Multi-path
- Increase capacity
- Less noise pollution

The current concept of MLS in MMR will not be capable of a multi-path approach and will not increase capacity, but can only match the capacity of the existing ILS Category III because of the lack of multi-path support. The current concept of MLS also will not decrease noise pollution because of the lack of multi-path support.

What, then, is the point of having MLS on board?

Well, it's needed to land the airplane in heavy mist three times a year. But if operators don't install MLS, they will save enough money to afford a couple of delays each year for the next 50 years. We earthlings had the courage to delay MLS for 20 or more years, but we don't have the courage to let it die. I now declare MLS dead. RIP MLS (Figure 8.2).

FIGURE 8.2   Rest in Peace MLS.

LTN-92: Carousel of Glass

The standard navigation equipment in the 747 classic was Delco Carousel IV. Carousel IV was a mechanical gyroscope (gyros). The technique was originally developed for the Apollo project to send astronauts to the surface of the moon. The spin-off was Carousel IV navigation equipment in the 747 classic. The heart of the system is an LRU called Carousel IV INU. There was a control panel installed at the cockpit ceiling panel to switch it on and off and initialize the INS. At the pedestal there were three control panels. The operation was relatively simple: the pilot or engineer would enter the starting location in latitude and longitude to initialize the INU. After initialization, it was switched to navigation mode, and a pilot or flight engineer would enter the individual waypoints (latitude and longitude). The system used mechanical gyros and proof-mass accelerometers to measure movement from the start point. Only nine waypoints could be inserted. Therefore, the crew was constantly entering the waypoints during the flight. BA later modified the system, installing the Flight Management System Multifunction Control Display Units (FMS MCDUs), which increased the number of waypoints, lowering the workload of the pilots.

After some time, Boeing started to deliver aircraft with PMS, (Performance Management System) which was a kind of mini-FMS. There was one more panel required in the forward pedestal, which is used for controlling the PMS. The pilots could insert up to 50 waypoints in the PMS. The PMS transferred those waypoints to INS, which steered the autopilot. This was quite a luxury solution for that time. Unfortunately, after 10 years of operation, the systems were not very reliable. Each aircraft had a triple INS system and a single PMS. Soon, no one was satisfied with the reliability of INS Carousel IV. The 2,500 hours MTBUR was a poor number.

I was a senior engineer within the avionics engineering group at KLM, and all fingers were pointed at me, requesting better reliability of the units. I organized a meeting with flight operations, engineering managers, and test pilots. My proposal was to remove Carousel IV and install Litton laser gyros (LTN-92). LTN-92 was a much better navigator, and it was already used on new 747s and in Lockheed C-130 aircraft. The certification was quite easy because there was an STC available, and Boeing was even willing to issue the Master Change for installation. There was no wiring change, just box swapping and pilot training, because the control panel was different. Early in the process, it became evident to me that pilots had a lot of power. I was counting on their acceptance of LTN-92 installation, but they immediately refused, because they didn't like the control panel. The second option was a different control panel, but that one was also rejected. Pilots wanted to get MCDU as in the 737-300 or DC-10 or 747-400. They didn't want just a simple drop-in replacement; they wanted an upgrade and a change of technology. I had a bad feeling that this would become a never-ending story. It was 1994, and aviation was on the verge of switching to new technology, but we stuck with that beautiful 747 classic and its old cockpit.

Canadian Marconi

Then I heard about Canadian Marconi and their new concept: FMC with GPS and MCDUs. I invited their European representative for a briefing. It was a strange meeting. The representative didn't bring anything to show or tell us. He just asked questions and took notes. Three weeks later, a few things happened. Canadian Marconi came back with a proposed solution, and another KLM engineer took on the project, because I had to finish my RVSM, SATCOM, and TCAS projects. History was repeating itself. Somebody else started the new project, and I was pushed aside, knowing that I would be involved again at a later date—when the project started to get difficult.

Due to the power of pilots and the influence of inexperienced engineers, the project started to grow out of control. The first rule of project management is: Make a reasonable plan. The second rule is: Stick to the plan. But we found ourselves in a situation where the plan was changing constantly, and the costs were climbing constantly as well.

After many meetings and trips to Montreal (to Canadian Marconi, which is now known under the name Esterline), the system was finally defined. It took at least two years to get this far. Instead of a Carousel IV, the LTN-92 laser gyros were installed. LTN-92 was connected to Marconi FMCs. Three new Marconi MCDUs were supposed to be installed in the cockpit at the pedestal. The pilots complained that they needed a display. From their point of view, the ADI and HSIs were not good enough anymore. New Liquid Crystal Displays (LCDs) should replace ADI and HSIs. Then the engineers responsible for the engines complained that engine instruments were not good enough, and they needed displays to present all engine parameters. So a display manufacturer agreed to do that, too. When we installed SATCOM in the 747 classics, we also installed a control panel for SATCOM called Multi Input Display Unit (MIDU). MIDU was installed at the pedestal, but there was a problem: The MCDU #3 had to fit in the pedestal, and therefore, the MIDU had to be moved from the flight engineer's panel. Now that was specified as well. To specify everything, and to contract multiple parties, took two years. It was a difficult, frustrating job. The Dutch are cheap, and the result of such an attitude was to prioritize a cost-effective solution and contract the cheapest suppliers for the STC data package and installation kits. I was happy now *not* to be involved at that stage of the project. One problem was that the engineers had to do the work to meet the schedule and didn't have much time for business trips and coordination meetings in Canada. Pilots involved in the project, on the other hand, simply asked to be scheduled to fly to Montreal. There they visited the OEM, had a meeting, and made the changes in the specification. The OEM was more than happy to deal with pilots than with engineers. Once the pilots decided on the configuration, the OEM informed the engineers that the specification had changed and that it was agreed with flight ops. The decision was made. Engineers were not powerful enough to argue with pilots and that was that.

Financially, the project was a disaster. The project engineer had to go to management with an additional proposal and beg for more money. The project became too big to fail: The contracts were signed and there was no escape.

Technically speaking, it was a revolutionary step. The 747-classic glass cockpit was born (Figure 9.1). It was already exciting, although physically, nothing had been done. On paper, it was an innovative concept. During those two years, I could only watch from a distance and answer the questions they asked, although the project team didn't like my answers and eventually stopped asking. The biggest problem was that there were six external parties (Kit supplier, Engineering data package supplier, and four OEM's), under contract as well as four internal parties (Flight ops, Material planning, Base maintenance, and Engineering). They blamed one another for delays all the time. And there were many delays. Eventually, it was decided to install more wiring than required. You never know when you might need extra, but 50% more wiring was definitely over the top. Then there was a delay in the delivery of the displays. After that, the data package was delayed. Finally, it was decided to split the project into two parts: part one, Laser gyros and FMS; and part two, displays. I guess that was the price of pioneering and doing such an extremely big project. It was perhaps too much for a single airline to take on.

FIGURE 9.1 Block diagram of 747-2/300 cockpit upgrade.

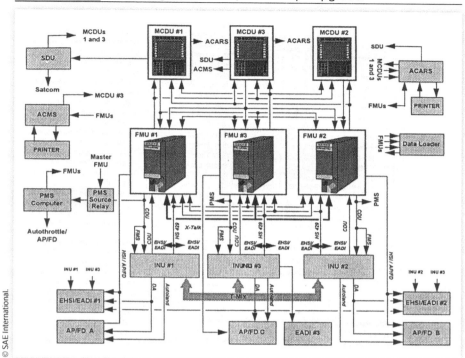

Part one was certified, and the entire fleet of sixteen 747s was modified. This was the easy part. The displays were not yet ready for certification. Soon, a new project appeared on the horizon: EGPWS. The manufacturer of 747 displays (Smiths) was asked if the displays could support EGPWS and the answer was: "No, but we are working on it because one other customer [TWA] wanted their displays with the EGPWS option for the MD-80 fleet, but that would be a different part number." The MD-80 aircraft, also known as Mad Dog, was an aircraft with classic instruments. For the Enhanced Ground Proximity Warning System on the Mad Dog, they decided to replace classic ADI and HSIs with LCDs. Those displays were a different p/n than we ordered for the 747 aircraft. They were able to present the Enhanced Ground Proximity Warning image. Our 747 p/n was not good enough to do so, but changing the contract was too expensive and modification of the ordered displays was not possible.

Finally, I completed my SATCOM, TCAS, and RVSM projects. I was spending my time running the avionics group and doing my senior engineer's work. One day, I was in the middle of the department with about 15 engineers around me. We used to have such meetings within the department rather than in the meeting room. We were sitting in a circle and discussing our work when the secretary of the Head of Base Maintenance came in and asked me to come out of the meeting because her boss wanted to talk to me. I said to the department team: "I'll be back." (In the well-known words and accent spoken by Arnold Schwarzenegger in the movie *Terminator*). I went to the office of the Head of Base Maintenance, and he said to me: "I have a project for you. It is the EGPWS installation on the whole fleet. I need a new person for that project because our current project director got a new job, and he is leaving next week." My answer was: "Yes, I can do that." "Do you want to talk to your wife first?" he asked, "or to think about that?" "No, it is OK. It is not necessary," was my answer. Then he said to keep it hush-hush and wait for further instructions. I returned to my meeting and finished it like nothing special had happened.

EGPWC Installation

The following Monday I was already in my new office. It was a big shock for the engineering department. I spent 14 years in engineering, and nobody thought that I would ever leave the department. My attraction to the new project was stronger than my nostalgia for the engineering department. It was an amazingly fast transition. If I had applied for a job within the company, it would take weeks to get the new assignment following a series of interviews. In this case, my position was changed in the blink of an eye. This had happened to me once before, when I was transferred from the calibration lab to engineering, and now it was happening again with a move from engineering to base maintenance. In my 14 years in engineering, I tried a few times to get another job following normal procedures and was always rejected. Now I didn't even have to apply for the job. Oddly, I started both jobs on April 1st, April Fool's Day.

There was something I didn't like about my job change: The person previously in the position had the title Modification

Director. When I took over his job, doing the same job he was doing, my title was Project Manager. I wondered why that was, but I guess we will never know.

I left the engineering department and my beloved avionics group. But the engineer in charge of the cockpit upgrade also left engineering. When the going gets tough, the tough get going. They also gave me the cockpit upgrade project to finish. The final, difficult part of the project was the installation of the displays, and the engineering team installing cockpit upgrades had already moved on to new jobs. I was the only one left with enough knowledge to manage the project to completion. I knew that the financial part would be difficult because of all the changes during the project. The person who had run the project before me told me that there was 2M USD still available for the project and there were no financial concerns. That was good news. After he left, I decided to investigate the financials for myself. The financial control group got all the figures together and announced that all the money allocated for installation was already spent. There was a zero balance, no money. Done. Nada.

Fortunately, the displays were paid for in advance. I only needed funds to pay all the work hours for installation and certification costs, which is about 1M USD for the fleet of now 14 aircraft. (Two aircraft had been sold to America West, and they were not my worry anymore). But how to get the necessary 1M USD? Stopping the project was not a good option. The OEM would deliver 50 new displays, and there would be no use for them. The 2M USD was already spent on the OEM, and they were set to deliver. My best option was to ask for more money to finish the project. It was a lousy situation. How could I say: Dear boss, I need 1M USD to finish the project? Otherwise, the 2M USD already spent was wasted on displays that might never be used.

It was just a one-page business case on my part. The amount of 1M USD is the final cost of the project. The pilots were supposed to provide the justification for the benefits, and they did: If they have a display in the cockpit with a map mode, they could save 1 minute on each flight. That was enough savings to justify funds to finish the project.

The proposal was eventually approved, and I had just one chance to complete the project. If the certification failed in the first attempt, the project would be terminated, and the costs would be booked as a loss. For now, I was back in business. All the fun stuff at the beginning of the project was completed by others, and the tough stuff was now on my plate. First, I killed the engine instruments emulation and engine displays. The OEM was not able to finish it on time and make it work, so I pulled the plug. Then I pulled the plug on the EGPWS option in the displays, which was optional. It would introduce additional costs, and there was no money for extras. I negotiated the certification plan with DERs and ground time for aircraft for modification. After that, I contacted the FAA test pilot, and we planned two test flights: One for proof of concept and check on the interfacing and one for STC. There were three months between those flights.

The first flight was flown by FAA test pilot Bill. He was the legendary pilot of the Lockheed SR-71. In 1966, he bailed out after his SR-71 disintegrated while flying at Mach 3.18 at 78,800 ft, and by some miracle, he made it down to New Mexico alive and well. I didn't know that I had such a famous pilot under contract. After I found out, I printed his picture from the Internet and asked for his autograph. It was amazing to even be in the vicinity of such a legendary pilot.

During that test flight, we had an unusual anomaly. Bill was executing a right turn, and on the display we could see an aircraft symbol turning left. It was not time to panic, but we were definitely confused. Somebody suggested calling Dr. MGS, the FMS father himself. He was the OEM FMS genius. We had a morning test flight but still decided to call him from the aircraft via SATCOM. In Montreal, it was nighttime. We woke him up and asked the question. He immediately answered: "Wire 9001-205-00-R and 9001-205-B were probably swapped. Swap them back, and you will be OK." Our engineer and tech went down to the avionic bay, swapped those two wires during the flight, and fixed the problem. Beautiful! That is what I call initiative. We had confidence in the system and in our people.

9/11

The second and final test flight was planned for September 2001. I hired a different test pilot (Figure 9.2). He was also an FAA DER

FIGURE 9.2 Excitement in the cockpit during the certification test flight, somewhere above North Sea on September 11, 2001.

Courtesy of Marijan Jozic.

pilot and had flown the 747 classic with winglets a few months earlier. He had lost one of those winglets in the Pacific; the second winglet used in that test program is on display at the entrance to the Museum of Flight in Seattle. In addition to the pilot, we had the management DER, EMI-DER, DAR (Designated Airworthiness Representative), the representative of the OEM and the representative of the display OEM and the engineer responsible for STC data package and drawings, our engineers, and a few technicians. It took us three days to do the installation and all ground and EMI testing. There was endless testing and then even more testing. Finally, the flight test was planned for the morning of September 11, 2001. The 747 was parked in front of Hangar 14, fueled, finished, and ready to fly. The crew arrived and the airplane departed. I was the only one left in front of Hangar 14. There was nothing for me to do but wait. The flight was planned for 4 hours, and it could even take longer. I had no choice but to be patient and hope that everything would work fine.

Four hours later, the 747 taxied in. The aluminum bird was parked in front of the hangar, and the crew, DERs and DARs, and engineers were directed to our big meeting room for debriefing and signing of the papers (Figure 9.3). The plan was to immediately de-modify the airplane and submit the papers to the FAA, and after the FAA processed the papers, we would modify the whole fleet. Our company would be the first operator flying 747-200s/747-300s with displays in the cockpit. Many airlines had already knocked at our door, and I again dreamed of establishing a modification team to complete this unique modification for other operators. After the installation of "our" cockpit upgrade, the 747 could fly another 10 to 15 years without any major concerns (Figure 9.4).

During that final meeting, everybody was in a kind of euphoric state. There was a lot of laughing and telling stories about the successful flight.

FIGURE 9.3   Aircraft parked after a test flight in front of the KLM H14.

Courtesy of Marijan Jozic.

FIGURE 9.4 747-classic P1 panel with new displays (ADI and HSI).

Courtesy of Marijan Jozic.

Just before I wrapped up the meeting, the secretary came to the room and asked me to come into the corridor. I did, and she told me that a Boeing had crashed into the World Trade Center (WTC) in New York, and the WTC was on fire. I thought it was probably something like that event when that military plane crashed into the Empire State Building in 1945. But this was something else. I went to my office and tried to get the news site on my computer, but it was not possible. The Internet was down, overloaded. I called my wife and asked her to switch on the TV, select CNN, and tell me what she saw. She described the burning tower, and as she watched, another airplane crashed into the second tower.

I went back to the meeting room and reported the bad news. We had such a successful morning, now followed by terrible disaster. People were silent, not knowing what to say. It was a very tense and sad moment. Late in the afternoon, many attendees had flights to catch back to the US and Canada. The flights were on schedule, and my certification team was hoping that they could be back in the US and Canada that day. We said our goodbyes and they departed for the airport.

Three or four hours later, my phone was ringing. Four Americans and one Canadian had returned to The Netherlands and needed help finding hotel rooms. The rest of the evening and night I drove them to hotels. The hotels around the airport were already fully booked, although the prices had gone sky-high. Hotels 50 or more miles away from the airport had rooms available, so we booked them there. It took three days to get them on flights home, and I was relieved when they called and said that they had arrived. What an end to a project.

Phase Out

Three weeks later, the management DER called to tell me that the FAA approved the modification, and I could plan an upgrade of the whole fleet. But by then, everything was different. The aviation industry had collapsed after 9/11, and there was not much flying. It was a true industry crisis. Discussions about the cockpit upgrade project started all over again. It was decided to go ahead and modify the fleet. Then it was decided to phase out the fleet and sell all 747-200/747-300 aircraft. Some people were afraid that nobody would buy them if they had a glass cockpit. So it might be better not to modify them after all. In the end, it was decided to modify them and sell them modified. If a customer preferred, we could modify the airplanes back to the original condition. To facilitate this possibility, we had to keep all the old components in the warehouse. One-and-a-half years later, the entire fleet was gone. The fantastic project, which took us 6 years and cost perhaps 40M USD, was forgotten. Everybody was assigned to a new project, and the cockpit upgrade was history. Or is it?

Over the next few months, after certification of the glass cockpit, I had been flying with our sales department to Hong Kong, Japan, and the UK, giving presentations about glass cockpits. Many airlines were making plans to buy our concept and let us do the modification on their fleet. Even Boeing wanted to cooperate with us and go on the market with our concept.

My business trip to Hong Kong was unusual. We arrived in Hong Kong in the morning after an 11-hour flight on a KLM aircraft. At about noon, I gave the presentation at Dragon Air in Dragon House (their headquarters [HQ]), and in the evening, I had an 11-hour Cathay Pacific flight back to Amsterdam. That was a lot of flying for just a 2-hour meeting.

The last aircraft from the KLM fleet was sold to Surinam Airways. This was a 747-300 known as Uniform Whiskey (because of its tail number UW). The glass cockpit was installed, but this was two years after certification, and I was deep into another project—the EGPWC. Our sales team told me to install EGPWS in Uniform Whiskey, because they had sold the aircraft and promised to deliver it with EGPWS. They started to pressure me again. This time I had no choice: EGPWS should be installed. And this time, I didn't have to deal with pilots. It was my call.

The installation was simple. Instead of the present weather radar PPI, I decided to install two new displays that can present the weather radar image and the EGPWS image. After some fiddling, the cockpit was completed: A beautiful glass cockpit with all the bells and whistles. It was the best and the most complete 747 classic cockpit ever. I was proud of it. It was my creation, my project, my masterpiece.

10

My Dream Team

There are a lot of theories on how to build a team. Hundreds of books are written about team builders, team managers, and leaders. I am convinced that many skills of team leadership can be learned, but there is also the *X* factor that some people have that can make them good team leaders. What is that *X* factor? I really don't know. But I know that I possess some skills that make it much easier for me to build a team.

Throughout my career, I have often had the misfortune to be surrounded by people whose ideas about working together differed from mine. Even worse, I have also found myself in a situation where people were dishonest or jealous of others' success. That attitude was not always professional, and it hurt the company we all worked for. When this occurred, I was sad but could not do much more than accept the situation and do my best to navigate around it. I realized that I had to perform better than anybody else in order to meet my targets. This was frustrating, because I knew there was an easier way, but I also knew that I had to exercise patience.

After many years of experience, I developed a normal pattern of operation. I knew that I would have to start the project and that, after some time working together, the team would start to cohere and then flourish. I didn't have to do anything special; just by being myself I could inspire a team to pull together.

TCAS Team

One of the first teams I led was when we were doing TCAS installations. After I released the documentation for modification, I just had to keep track of the flow of events. Simply showing my face at kit planning was enough to let the team know that I cared about all the work they were doing. This showed the team my positive attitude and helped us develop friendships. If there was a problem, they would know they could call and ask me for help. The team, and not only the team, but every team member, had confidence that I could solve their problems. When they called for help, I responded quickly, and I made sure that any necessary materials were ordered and delivered before the airplane came to the hangar.

If there was a dedicated modification line, the work could be divided into a few steps. First, I made sure that all preparations were done. Good preparation makes the operation of the whole team easier and run more smoothly. Normally, the preparation started with parking the aircraft in the hangar, getting all materials on deck, and making sure that all hatches and panels were open. Then the power was removed from the aircraft. The next few days were important. The team worked on installing the wiring and components. For that part of the job, I didn't need many people, just a small group of specialists. After a few days, the installation was done, and the aircraft could get electrical power again. This was when the testing and troubleshooting started. After that was done, the aircraft was ready for delivery.

In the just-described process, there are a couple of important things. The project manager should visit the site very often, not only to show his/her face but also to listen to the technicians. Listening is the most important part of my job. And not just listening but making sure the team feels that the manager understands their problems. Once the manager develops a real connection with the team, the team will come to him/her not just with problems, but also with suggestions, because they believe that the manager can help them and make their life easier. This is the real secret to good management. After the manager gives the staff that feeling, the relationship can only become better, and love will start to grow between the team and the project manager. Yes, love: It is about love.

Here are some of my team's suggestions after the first installation of TCAS:

- Open only a certain area of the aircraft: Some hatches don't have to be opened.

- After the power is removed from the aircraft, there is a group of four people per shift not doing much except waiting for the specialists to finish their work so that they can start to close the hatches and panels when the installation team is done. They sometimes ask if there are complaints in the cabin maintenance log. They will be able to repair seats in the three or four shifts while they are waiting.

- The need for an accurate list of point-to-point measurements should be made available to measure all new wiring that is installed. This is easier than working from a schematic.

- Staff asked for a list of hot pins to use to measure whether the power is at the correct pins before they insert the LRUs. This can prevent a short circuit or damage to the LRUs.

- Team members also asked for the list of positions of all switches in the cockpit in order to deliver the aircraft in perfect order.

Some of these things were supposed to be delivered in the first place. The project manager should make sure that these requests are completed before the next aircraft arrives. If the staff requests are satisfied, you will have a happy team. This experience was a big learning moment for me as a project manager.

SATCOM

A few years later, a new project was running: SATCOM. As the project manager, I had to be there for the entire first installation and observe the full process. I talked to the team, learned from the first installation, and carried that knowledge through to the next installation. The team members on the work floor (hangar) will always figure out a better way to do something. If you make it easier for them, you will have less to worry about or fix later.

SATCOM was a new and different system than the team was used to dealing with. The only way to do the necessary trouble-shooting was to use a laptop. The SATCOM engineer needed a laptop and a cable to connect the laptop to the SATCOM SDU (Figure 10.1). In those days, laptops were a luxury. Only high-ranking managers had a laptop. It was something of a status symbol: If you have a laptop, you must be important. But for a SATCOM engineer, a laptop was a standard tool. I had a good relationship with my team, and they built me a cable to connect my laptop to the SDU. They came to me with the idea of making a power cable to connect the laptop power cord to the connector of the Hot Cup in the galley. The galley was a good place to be when trouble-shooting the new installation, because in the galley there was a hatch that could be opened, and there, just above the ceiling, was the SATCOM rack. If I climbed the galley desk I could easily find the connector of the SDU and connect it to the laptop. Then I would run the PROCOM software and test the SATCOM system to get a

FIGURE 10.1 You cannot troubleshoot SATCOM without a laptop.

guteksk7/Shutterstock.com.

list of all the failures. The SDU computer would troubleshoot the whole system!

At one point, there was a software modification for SATCOM, and the new software -20050 had to be installed. One part of the installation was to insert additional data into the SDU; this was something like installation date and place. It was part of SB, so we had to do it whether we thought it necessary or not. To make my team's life easier, I trained them to do the modification and gave them my laptop. Well, this was a major event. They were happy that they got the laptop and the responsibility, but management was angry that I gave an "expensive" laptop to the technicians. The "expensive" laptop cost no more than 1,000 USD at the time, and the team was working on SATCOM components that cost at least 100 K USD each. I trusted my team, and I knew that they would not break the laptop. And, of course, they didn't.

AMP

A few years later, there was a well-known, huge modification campaign: Avionics Master Plan (AMP). I knew from the beginning that it would be big, and I wanted to prepare the staff for such a large project. I decided to organize a workshop a few months before the first aircraft would be modified. Many managers complained that I invited their staff to a half-day meeting (workshop). There was a lot of opposition, but I had to prepare them for this project. Leadership had to look at the big picture. There was something totally new planned to be installed in the aircraft. By the time they learned about it in a standard refresh course for their license, they would have already faced it in the hangar during the installation. That workshop was crucial, and the team was happy to have had it. They told me later that it was the first time in their career that somebody had organized such an event before they started the modifications. At the end of the modification campaign, I decided to celebrate the successful completion of the project and invited my team to a restaurant for a party. They did a great job, and the whole project was completed within budget and on time. We all had reason to celebrate.

Modification Team

My team was superb, and they were working on the first line of new technology in the transport aircraft. I thought we should sell the knowledge, experience, and skill set of the team. This was the 1990s and 2000s; I was already known as a good project manager, and I was familiar with the many modifications on the market. My idea was to develop a modification department at KLM. Lufthansa already had a solid group that specialized in modifications. My team was just as good at aircraft modifications, and I had some excellent engineers, too. There was a lot of interest in such a service in the market. Unfortunately, internally, there was a lot of opposition to the idea. The only opportunity for the team was to do some installations abroad. Organizing this was a thankless job, because the managers who were supposed to assign the staff to such projects didn't want to cooperate. Only with a lot of negotiation and escalation could I get two or three people to travel to the customer's site to help accomplish the modification. I was never allowed to go, which was very frustrating. Often, I had to be available day and night just in case the onsite team needed my assistance. If you want to destroy a team's morale, this is a good plan: Kill any initiative and keep staff constrained.

Boeing Partners

I also showed that I could build a team with other organizations. In the 1990s, a cargo operator purchased four 747 classic aircraft from a Thai airline. Those aircraft had an electric airspeed indicator that was unreliable. This was when I was just a systems engineer. I was asked to replace the indicators with pneumatic airspeed indicators. The work had to be completed within 10 days, and the installation was supposed to take place in Wichita, Kansas. I issued the engineering order and sent the list of materials needed to the staff in Wichita. Boeing asked me to come over to Wichita for the first installation. Everything was lined up, and I was at the site. The Boeing technicians were happy to have me there, because

I could tell them what to do and update the drawing if there was a minor change. Because I was there day and night, they started to appreciate my presence, and we became friends.

Everything was smooth sailing until we powered the aircraft and found that the autothrottle bug on the indicator was not working. I had a difficult time figuring out what the problem was. A new friend, a Boeing engineer, called me at the hotel and told me that there must be an interface problem between the indicator and the mode select panel. I called my team in Amsterdam, and they investigated the issue and came back with an idea to change the indicators. The new part number should be -013 instead -016. I didn't sleep much that night. The next day I told the operator that I needed new indicators (Figure 10.2). I was prepared to get a lot of criticism because the two new indicators cost 35K USD each. But the aircraft owner just ordered them and told me that they would be shipped on a special flight from Miami. Can you believe it? No discussion over cost. Toto, we're not in Amsterdam any more!

FIGURE 10.2 747-1/2/300 Mach Airspeed Indicator (MASI). On the left is the pneumatic and the right is the electric indicator.

The indicators arrived, and after we installed them, everything was fine. I fixed the paperwork and got a new 8110-3 from DER so that everybody was happy. The team worked perfectly: We worked together toward a common goal, and we claimed victory. Perhaps this was in part because we all were speaking the same engineering language, and we were working on the same challenges. This story is not finished: The cargo operator didn't want to release me from duty. I had to stay until the flight test, after maintenance and conversion of the aircraft to freighter configuration was done. The next five days went as follows: I would check out from my hotel in the morning, hoping that the aircraft would go on the test flight that day. If there was no test flight, I would come back to the hotel and check back in. The next day followed the same pattern. After three days, the hotel staff started to call me Mr. Checkout, and they gave me the same room back every day after that. After five days, we had the test flight, and when the operator was convinced that the system was working OK, I was released from duty.

Approximately one year after that event, my company sent the 747 airplane to Wichita for conversion to a freighter. I was hoping to be able to travel to Wichita and be there during the conversion as a representative. According to planning, two aircraft would undergo the conversion, and I would be there for three to four months. But no—there was some office politics, and I was told that they needed me in Amsterdam. Later, I learned that somebody unknown to aircraft maintenance would go in my place. So be it. A week or two after the aircraft was parked in the hangar in Wichita, I was told that there was a SATCOM problem with the aircraft and that I must depart for Wichita. A week later, I entered the hangar in Wichita. It was a well-known hangar where they built the B-29 bombers in the Second World War. The hangar was not high enough for the 747s, and therefore, they first removed the tail (vertical stabilizer) and then parked the bird in the hangar. That was a very strange scene, a 747 without a tail parked in the hangar.

One of the people in the hangar recognized me and came over to shake hands. He said: "We asked for Jesus, and we got Jozic!" They were obviously happy to see me. The representative we had sent was not technically skilled, and the Wichita team couldn't communicate very well with him. Therefore, they asked for me. The whole group of Boeing technicians greeted me enthusiastically. Then I learned what the problems were. The biggest issue was the SATCOM wiring at the main deck. The wiring was installed under the ceiling. That was OK in passenger configuration but not in cargo configuration. Because the wiring stood in the way when the containers were loaded on the main deck, they had to remove it. Therefore, the SATCOM was not working. They had just cut the wires somewhere by Door 11 and by the STA-1230, which was halfway off the main deck. My job was to fix that problem; not too difficult. I was able to do the drawings and the entire engineering order on the spot. I knew the Boeing engineers, they were the same guys who were in Amsterdam for the Martin Air project (see Chapter 5: The Last 747 Classic). Now they helped me with some documentation. My friends in the hangar helped me with getting materials from local stock, and the problem was solved. Now I had to do something to help them. Since I was "The guy" (that was what they called me), I got a list of issues they had provided to our representative several weeks ago that were supposed to be resolved quickly. It was a simple thing, but KLM's representative was not even able to understand the questions.

This experience demonstrates how it works with a team—when the team is working well. It is give and take. In the end, everybody was satisfied, and the airplane was delivered on time. They had helped me with SATCOM wiring, and now it was my turn to help them with those pesky issues that should be resolved during the airplane's ground time.

I also trained the Boeing team to troubleshoot the SATCOM system using my laptop. Our aircraft was the first with SATCOM in its hangar. This was new technology for them, and they were happy to be able to talk to an experienced SATCOM engineer. It was also very rewarding for me.

Dream Team

To recap, the team is essential. If a project leader gives their best to the team, he or she will get the best performance from the team. I have seen many project managers who were convinced that they didn't have to know the technical details about the project because the team should know it. Well, it does not work that way in real life. The success of the project is 80% dependent on the project leader. If he or she does their part and stays open and ready for the team, success will be guaranteed. Forging meaningful relationships with your team members will make you the best project leader. As you can see, I am carefully saying *leader* and not *manager*. A *leader* is one who leads, inspires, motivates, encourages, and influences the team to work toward the organization's objectives. *Managers* are people employed by the organization with the goal to direct and monitor the work of other employees working in the organization. They have the authority to hire or fire the employees.

I was always pleased when people called my team "The Dream Team." It was very rewarding for me to hear, and it proved that I was doing my job in the best possible way. I was always proud of my team.

Can you imagine how I felt when my manager said to me after I finished the project AMP and successfully modified 100 aircraft in a period of two years: "We have done all the modifications. We don't need you anymore!" I couldn't believe what I had just heard—not words of appreciation, but a dismissal. One week later, I was on a new job and had to start building a new dream team. First, I had a dream team in the engineering department. Then I left them for a new job at base maintenance. There I created a new dream team, and after a few successful modification campaigns, I had to go. I became the head of the shop engineering department. The staff I met there was not yet a team at all. They didn't even talk to one another: Twenty-four people were sitting in the office of a shop engineering department, turning their backs on one another, and not knowing what their direct colleague was working on. I had to start all over again, but this time, at least, I knew how to build the team. The irony was that after 14 years, and after I did such good work with my team in the shop engineering department, I was told that I had to leave because, after reorganization, I was not needed there anymore. This is often the fate of the manager who knows too much about airplanes. Sometimes they are seen as a rebel and they have to leave.

11

AMP: Avionics Master Plan

Statistics show that due to TCAS, air traffic collisions have been nearly 100% eliminated. This is great news. The next big cause of aircraft crashes is CFIT: Controlled Flight Into Terrain. CFIT is an accident in which an airworthy aircraft under pilot control is unintentionally flown into the ground, a mountain, a body of water, or another obstacle. In a typical CFIT scenario, the crew is unaware of the impending disaster until it is too late. The term was first used by engineers at Boeing in the late 1970s. Halfway through the 1990s, the idea of a Ground Proximity Warning System (GPWS) started to take shape. The earth had been fully photographed by satellites, and there was a fairly thorough and accurate database of the landscape. All mountains and valleys were in the database. Computers were more sophisticated by then, and there was plenty of memory for storage. The time was mature to do big things, and the next big thing was EGPWS (Enhanced Ground Proximity Warning System). Many aircraft were equipped with GPWS, which was quite a primitive system and not adopted worldwide because it simply wasn't good enough. FAA mandated an improved system called EGPWS on March 29, 2002. Just for clarification: EGPWS is the name given by the OEM. The FAA introduced the generic term TAWS (Terrain Awareness and Warning System) to encompass all terrain-avoidance systems that meet the relevant FAA standards, which include GPWS, EGPWS and any future system that might replace them.

EGPWS Modification

Back in the late 1990s, engineers were faced with a dilemma, because there were many options to comply with that mandate. Also, it was not certain that GPS was needed, because laser gyros with DME-DME updates were quite accurate. The options included:

- Connect EGPWC to laser gyros and present the image on the display. The voice callout is connected to the same speaker as TCAS.

- Stand-alone GPS receiver to provide the GPS position to the EGPWC.

- Install EGPWC with a built-in GPS cart inside the EGPWC.

- Install MMRs with ILS and GPS, and provide the GPS information to the EGPWC.

The situation differed based on aircraft type, and the financial calculation of the aircraft system could be complicated for companies with several aircraft types in their fleet.

An engineer might also be concerned about the certification. The engineer could choose the STC or Master Change. There is a big difference between the two, cost- and risk-wise. STC is a major modification process by the FAA. The STC owner takes in all risk for the installation. The risk is manageable if you know what you're doing. The biggest problem would be that, after installation, should the system not work for any reason, you must accept a delay and try the process again. Most of the work is in the hands of the engineer (materials, drawings, certification plan, testing, etc.). It is very exciting work, and if done properly, it is also quite rewarding. Unfortunately, sometimes the costs will be higher than predicted because of delays. The company must also pay a significant amount of money to the airframer for the update of the Wiring Diagram Manual (WDM), Illustrated

Partlist Catalog (IPC), Aircraft Maintenance Manual (AMM), and Hookup List.

If the engineer decides to go for Master Change, all risks (technical and financial) will be in the hands of Boeing or Airbus (the airframer). The airframer provides a drawings package, FAA and EASA SB, and the kit and updates all books (AMM, IPC, WDM, and Hookup List). The operator purchases the LRUs called out in the SB and does the installation. This process is approximately three to four times more expensive than STC, but all the risk falls on the airframer, which can be a reassuring option.

In the second half of the 1990s, I was the engineer in the position to make these decisions. There were a lot of discussions and calculations used to pick and choose the right configuration and certification path. I had an option in mind that was feasible and doable. This choice also allowed for future growth. It was a viable option.

At this exact time, the MLS project was canceled, and our pilots were angry that they didn't get their new toy. Actually, the pilots didn't care, but there was an office at flight ops that always came up with new cool stuff. They lost the MLS battle, and now it was time for their payback. My boss, the Vice President (VP) of Base Maintenance, took me to a meeting with the CEO of Engineering and Maintenance, Head of the Engineering Department, and VP of flight ops. The VP of flight ops had been a 747-test pilot, and I knew him very well. I was prepared to explain my plan for EGPWS, but my boss told me to stay quiet and listen. I was not allowed to speak.

We entered the meeting room, and the pilot started, "We decided not to install the MLS in our fleet. But I can't send my boys [pilots] on a flight without GPS. When you install the EGPWC, make sure that GPS will be the source for the aircraft position. Can you do that?" and he looked at me. I said, "Yes!" At that moment, I realized that we had just made the decision to spend 55M USD. The meeting lasted no longer than 15 minutes, and I got myself a full-blown project. For the first time, I had the opportunity to start the project, and with a bit of luck, I would finish it. I was convinced that it would be a good project for me.

Planning

There were some new engineers assigned to the project. They had never worked on a large project, so some guidance was required. Those engineers could be compared to American football players. When the football player gets the ball, he holds the ball under his arm and starts to run like his life depends on direction and speed. This was my engineers. They would start to develop something and were not amused when, just as they got running, I called them back and made them restart from scratch. The first step was to make a reasonable plan per aircraft type. It was not that easy to get this done because the managers were pushing them to do something and show the results. After some negotiation, I got six specification reports for six aircraft types in the fleet: 747-200/747-300, 747-400, MD-11, 767, 737 classic, and 737-NG.

It was obvious that some aircraft could be modified during C and D checks, but some should be modified in a dedicated modification line. I had to calculate all the work hours needed for technicians, ground time and hangar space, kit planning effort, and so on. Based on the specification, I had to go to Boeing with a few of my engineers to agree on which SBs should be accomplished and what Boeing's part of the project would be. In total, we agreed that Boeing would provide about 20 master changes. A big concern was that, if we missed something, one of the aircraft types might need more rework. The full business plan was submitted to the fleet services department for approval. They had never had such a big project, but after a few sessions and an explanation of the scope of the project, they approved the project with the exception of 747-200/747-300. That aircraft was planned to leave the fleet before the FAA's EGPWS deadline. It would no longer be the fleet services department's problem, but I knew that I would see the aircraft again, and I had a plan: A poor man's solution for the 747-200/747-/300. See Table 11.1 to review the modifications per aircraft type. There was a lot of work to be managed.

TABLE 11.1 Systems and components affected by the EGPWS modifications.

Systems	747-400	747-200/ 747-300	767	MD-11	737-PG	737-NG
EGPWS	X	*X*	X	X	X	
MMR retrofit			X	X	X	
FMS retrofit			X			
EFIS Ctrl. Panel	X		X		X	
GPS Ant.	X		X	X	X	
EFIS mod						
MMR GPS mod.	X					
FMS mod.	X		X			
EGPWS options mod.	X	*X*	X	X	X	X
Master caution					X	
ODFDMU		*X*	X	X	X	X
FDR	X		X			
EFIS SGU			X		X	
Provisions dual FMS					X	
DFDAC	X					
EICAS OPC + mod.			X			
ADL			X			
CAWU						
CFDIU/OMT				X		
ACMS	X		X	X	X	X

We started with just a few engineers, but more than 500 people with various skill sets were involved over the course of the project, which became known as Avionics Master Plan (AMP). I purchased 500 red lanyards with the text AMP, so that everyone involved in the project would have one. In the end, I had only five or six left. They are now a collector's item. Many years after the project was completed, I used to run into people with that lanyard around their neck. Management didn't like it because red was not the brand color. The brand color was blue, and I had ordered red lanyards. They complained about this only once; because the project was running smoothly and within the budget, they refrained from further grumbling.

The most expensive part of the fleet to modify was the 767 fleet because Pegasus FMC had to be installed. Pegasus was able to process the GPS data needed for EGPWC, so, the pilots got a new toy, the sophisticated Pegasus FMC.

The most complicated aircraft to modify was a 737-PG, which some called the 737 classic. It is a small aircraft, and a lot of effort and skill was required to install the wiring and new components. The 737 was also the only type of aircraft that had to undergo the certification test flight.

Preparation Phase

After the specification and budget were approved, I could feel that the project switched to the next phase. The next phase was the preparation phase. This is an essential phase. If I prepare the project well, the actual project, the aircraft modification, is more likely to go smoothly.

The next item on my agenda was to negotiate all required contracts. I already had all the proposals ready in order to get the business case submitted and approved. Money was available for materials, SBs, man-hours in shops, work hours in hangars, the data package for flight simulators modification, and approximately 1.5M USD for avionics shop test equipment. There was a project buyer assigned to do all the purchasing work for me, but I had to attend each meeting with suppliers. I came to realize the importance of having a good contract. I learned one important rule: I can't unsign the contract; therefore, I should think twice before I sign it. We started with proposals, and the target was to lower the actual price before we signed the agreements with the parties.

If at all possible, such negotiations should not be done under pressure. In my experience, pressure often results in a bad deal for at least one party. I strive for a win-win situation. I know that I can squeeze the supplier until they start to cry, but this is likely to backfire later when I need something from them. To settle the score, whatever I needed would cost a lot or be delayed, and if I needed it sooner, I would have to pay the top price. A win-win is the best outcome for any deal.

I had some difficulty within my internal departments. Many years later, I learned that there are special courses for negotiation within a company. Unfortunately, I had to learn how to do this the hard way. For example, the team from the flight simulator asked me to negotiate the interface control documents for the units they had to modify/replace in their simulator. I asked them if they knew the price. The answer was, "Yes, we have experience with that. Normally we pay 250,000 USD for the data package. OK, the math would make that five simulators for 1.25M USD. However, if I negotiated a better deal, I could still bill the flight simulator department 250K per simulator. It was my budget, and I was planning to send them a bill via the internal payment method. But in our negotiation with suppliers, we got the whole data package for the flight simulator for free. It was the easiest item for suppliers to give away for free, because they didn't have to lower the price of hardware or SB. The interface control document was in their library, and it would not cost them more: The costs of developing it were already spent. My flight simulator department didn't like my methods at all. They escalated the issue to top management, who decided to slash those costs from my budget.

A second example was my avionics shop. They were allotted 1.5M USD for test equipment, but they didn't know what to do with the funds. Their controller called me and asked, "There is 1.5 million on the balance sheet, what shall we do with it?" My answer was, "If you don't know, just transfer it to my private account." The controller was angry at my response, so he called my boss and asked the same question. My boss said: "If you don't know, just transfer it to my private account." Now the controller was furious. We had to report to the CEO's office. He said: "Ha, ha, this is funny. But don't do it again." Eventually, the shop was able to purchase the IRIS 2000 test set and the TPS for EGPWC. So the 1.5M USD was very well spent (Figure 11.1).

During the preparation phase, we made several trips to Seattle and Long Beach. Over the years, I have met many excellent Boeing engineers and project managers. After I connected with them for this project, it was fun to work together. Boeing and Douglas had

FIGURE 11.1 IRIS 2000 ATE.

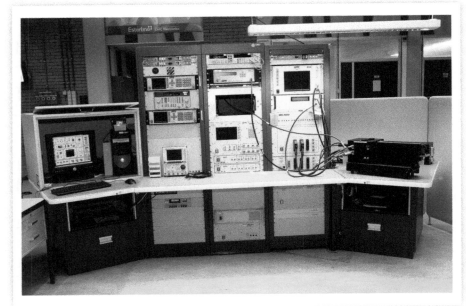

Courtesy of Marijan Jozic.

merged, and they were in the process of integration. It was difficult for them to keep track of our joint project and all its details. After spending time together, we gelled as a team. It was no longer their part of the project or our part of the project. It was all *our* project, and we collaborated fully.

The only difficulty in the engineering part of the project was when, at the very last moment, Boeing did the integration test of the 737 system, and in the lab, they discovered that the control panel had an EMI problem. The EMI was causing the LCD to blink every time the pilot changed the frequency. This was discovered near Christmas. We discussed the situation, called the OEM, and eventually agreed on how to fix it. The problem then became time. The fix had to be done fast, because we had deadlines to meet. The OEM decided to get a few technicians to work on Christmas Eve and modify the panel. The panel was tested satisfactory, and the SB for the panel was issued. Boeing was able to submit the correct control panel p/n and modification status to the FAA. Everything was timed perfectly, and we managed another crisis.

Implementation

In January 2003, the implementation phase of the project began. I felt in my whole body that a new phase of the project had started. This is difficult to describe, but for me, it is an emotional moment. I had a team in the engineering and modification preparation department, and now I had to build a team of technicians in the hangars. This was not difficult. I had a good relationship with that staff, many of whom worked on the TCAS and SATCOM modifications. They knew that I was the kind of manager who was always there for them, and I knew that they would be there for me as well. My problem with this team was their management: They were uncooperative.

Project managers worldwide know this feeling. Normal organizations always have problems with project organization, and project organization always has problems with normal organizations. Negotiations are required. There were plenty of technicians who wanted to work on my team. I had to build the project group, and I was dependent on technicians normally doing C or D checks and only small modifications. Now I had 2,000 work hours scheduled to complete the modifications in just one 747-400. Because it had to be done in one week, a dedicated group of technicians was required for my project, and I was dependent on the managers of the C or D checks for this staff. The managers were playing an unprofessional game. They would assign ten people per shift for my project, but when they were supposed to show up, I learned that they were on holiday or ill or also assigned to another project. Because of this, I started the modification with five or six people instead of ten. This caused me to be consistently late with aircraft delivery.

Another trick the managers tried was assigning a different group of technicians for every new tail number. In every big modification, there is a learning curve. For the first aircraft I needed 2,000 hours, for the second 1,500, and for the third and fourth 1,250. This was an optimum result for a good working team. When the managers assigned a brand-new set of technicians to my project,

I went back to needing 2,000 work hours. By the time we were working on the final airplane, everyone had experience on the project, and 1,250 work hours was a standard number. And then I didn't need them anymore, because the fleet was modified. Such frustrations followed me for all two years of the modification process.

I spent unnecessary time in those negotiations with hangar managers. At the same time, there were many small problems in the shops, kit planning, modification campaigns, and many other things. Once, I met with a good friend from kit planning who was now working at material services as a planner. She had been my kit planner during the TCAS project. I was feeling desperate, because the modification of the 767 Pegasus FMC was spiraling out of control. I went to material services to see if there was something we could do about this, and I ran into her. She was the type of person who was always dedicated to doing the job in the best possible way. My work with her followed a pattern: First, she would argue with me and give me a hard time, then she would argue with everybody else, and then she would buckle down and help me. This happened in the same sequence every time I dealt with her. She was one of a kind. Eventually, I was happy, she was happy, and the project went smoothly. Her job was to organize all materials for C and D checks. She added my modifications to her package and made it happen all over again. For two years, we didn't have a single problem with materials. She deserves a huge thank you.

The AMP project was a success. Then our sales team learned about the success of the project and started to investigate whether it might be possible to sell it to a third party. At that time, we had approximately 35% of the work in the hangar for a third party. The hangar management was only interested in selling work hours of sheet metal work, not modifications. It was too complicated for them, and only a small group of technicians was ever involved in such a job. Our first project for a third party was for Air France. Sales subcontracted four 747-400s. It was a good price, and my techs were excellent. I agreed with Air France to ferry the aircraft to Amsterdam on Sunday afternoon. A small team was present, and they parked the aircraft in the hangar, opened the ceiling at the upper deck and left-hand side of the airplane, and removed some LRUs. This way, the team that showed up on Monday at six in the morning could immediately get to work on the wiring. The night shifts were for the wire wrapping team. By Wednesday, all wiring was installed, and we could begin testing and closing the hatches. By Friday morning, the aircraft was done. I only had to call the tank crew to pump 25 tons of fuel into the aircraft and get the crew to pick it up. The same process was followed for the next three weeks, and just like that, the first third-party project was completed.

My next trip was to South Africa, to Nationwide Airlines. They had one 767, which was flying between Johannesburg and London. In Johannesburg, I did the presentation, and we immediately signed the contract. The salesperson was flabbergasted that it went so fast. Unfortunately, my management let me down with that job. The contract was for C check and EGPWS modification. We had 10 days of ground time, which was more than enough for both jobs. But they didn't start to work on the aircraft for 4 days! The customer was unhappy. The airplane was supposed to be ready by Friday, fly to London, and, by Sunday, take 260 people on a flight to Johannesburg. We delivered the aircraft on Monday morning. As a consequence, the airline had to put all the passengers up in a hotel for one night. As compensation for this, we didn't charge them for the modification. I was disappointed with my company's attitude, and I thought my days as modification manager were at an end. But management didn't let me go. I knew too much.

In 2005 the AMP project was finished on time and within budget. Some people complained that I had purposely provided a much too high-cost estimate in the beginning so that it was easy to finish at 8M USD below the budget. From this, I learned that I should spend the full budget next time. Management seemed to prefer when projects were failing; they would have to deal with it if the costs escalated.

The last phase of the AMP project was evaluation. That was when everything was supposed to be finished, and we discussed the project process in order to learn from it for future projects. The organization should learn something whether or not the project is successful. In the case of AMP, no one was interested in hearing about what went well and what went wrong. The project was terminated on February 1, 2005. I learned a lot about the organization, managers, myself, and about managing external engineers (Boeing, Collins, Honeywell, ECS) and procedures as the AMP project manager. I cherish the good friendships I made at Boeing, Honeywell, and Collins. Even after 20 years, whenever we meet, we reminisce about AMP.

FIGURE 11.2   Modern aircraft are delivered with TCAS, SATCOM, GPS, and EGPWS; no retrofits are required.

Courtesy of Marijan Jozic.

12

Stay Focused!

August 2004 was a warm month in Europe. I was working a lot that spring, but my employer didn't want to pay me for all my hours. As a form of compensation, they offered me free time. For every hour of overtime I worked, I got 1.24 hours off. There was one more company regulation: During the summer months, I was not allowed to take more than two weeks off at a time. As you know, engineers are used to working in extraordinary situations, and therefore, we are adept at finding extraordinary solutions to problems. I took my family by car to our summer house in Croatia on the Adriatic coast and stayed there for two weeks. Then I returned to The Netherlands and went to work the next day. Business as usual. Then I asked for another two-week holiday and went back to Croatia to my family for two more weeks.

I was not used to having so many days off, and I noticed that my family was becoming bored on such a long holiday. We decided to close the summer house and drive back to The Netherlands. It was a two-day drive, but we were on holiday and decided to take our time and enjoy the trip. After 5 hours of driving, we were on the German highway between Salzburg and Munich. In 3 or 4 hours more, we would reach our goal for that day. My mobile phone rang, and on the phone was Bastian, my friend from sales. He asked me when I was coming back to work, because they needed me for a special job. It was Wednesday, and I told him I wouldn't be back in the office before Friday. "OK," he said. "I will take care of the details. Come over to my office on Friday. The tickets will be ready, and you can go to Fort Lauderdale this coming weekend." What? Apparently, I was going from Croatia all the way to Florida with a layover in The Netherlands.

This was my assignment: My company wanted to contract two 747 cargo aircraft for D checks. The catch was that we were to upgrade those two aircraft with "just a few" modifications. They would tell me the details after I arrived in Fort Lauderdale. I was supposed to meet with McElroy, the CEO. They call him Mac. No other details were offered. Except that—oh, yes: Those two aircraft have been parked in Roswell, New Mexico for six years. They were previously owned by a cargo operator. And by the way, these are 747 classics. Well, enough already! They were lucky that I had the guts to open this Pandora's box. There was a D check planner traveling with me who was responsible for planning the D check and all job cards. But the project depended on me. I had to complete the contract for the modification job. Which modifications? Mac would tell me on Monday.

Focus HQ

Picture it: Monday morning, Fort Lauderdale, Sheraton Hotel at Commercial Boulevard. I started my day with an overdose of sugar in the form of a typical American breakfast of pancakes and maple syrup. Unhealthy, but I liked it. "Let's go, Marcel [the planner]. On the way to Focus HQ." Focus HQ was not far from Fort Lauderdale Airport. The GPS brought us to a parking lot under the highway. "Your destination is on the left," said the GPS. There were some shipping container offices but no signage, and nothing that looked like an airline headquarters. We asked a man in the parking lot, and he told us that we were at Focus, and Mac's office was on the first floor.

Well, this didn't look promising. Upstairs was an ordinary waiting room like at the dentist, with a few chairs and a shelf holding some old magazines. There was a silent guy sitting in the corner. Marcel and I took the other two seats and waited in silence. You know the feeling—it also happens in the men's room: Nobody talking, everyone minding their own business, no eye contact.

After a short wait, Mac called us into his office. Immediately, I could tell I was with the boss. Mac was straightforward: "I want those two aircraft to be modified to the latest standard: TAWS, ELT, ELS, LTN-92, GPS, ACARS, Iridium, and 8.33 channel spacing, and if there is something more, let me know. Laurine [the silent guy from the waiting room] is FAA DER. He can help you. I need your proposal within four weeks. Any questions?"

"No, Mr. Mac!" I said. In less than an hour, we were back in the parking lot discussing the project with Laurine. Mac was a tough, no BS kind of guy. And he obviously wanted decent aircraft.

Some background on Mac: Mac established a new cargo airline. He purchased two old 747s to fly cargo and he was planning to buy two more 747s. The new aircraft should get D checks and be modified to the latest configuration to enable them to operate without problems for the next six years, until the next D check. My job was to be the modification manager for those modifications and coordinate it with the completion of D checks. It looked like a complicated project. At least Laurine seemed quite experienced. He used to work for Eastern Airlines, and now he was an independent DER often contracted by modification companies. Laurine would make a good partner for me on this Focus job. Many years later, I learned that Laurine had been the Chairman of the ARINC AMC conference between 1974 and 1976. In that regard, we had something in common, because I was Chairman of the ARINC AMC conference from 2012 to 2019.

All documentation (in 2004, microfilms) was already sent to my office. The birds were now operating for a cargo operator and had been parked for five to six years. Is the documentation correct? That was my first worry. To save costs, some operators don't spend money to update their books. They might keep a box with temporary AMM and WDM pages rather than spend tons of cash on updating AMM and WDM. I had worked on such cases in the past. I crossed my fingers, hoping that this time, the microfilms were correct.

There was no time to waste in Florida. We took time to have dinner in one of the local restaurants (Figure 12.1). I tried alligator meat and learned never again to order Key West shrimp: It was too much work to peel them, and your hands get very dirty. For 50 cents more, I could have had shrimp that were cleaned and ready to eat. At least I can say that I ate gator meat.

FIGURE 12.1   In Fort Lauderdale, eating shrimp the Key West way and discussing the Focus Air Cargo project.

Courtesy of Marijan Jozic.

I took a direct flight from Miami to Amsterdam along the north side of the Bermuda Triangle back to my office to get started on the project (Figure 12.2).

FIGURE 12.2 On my way to and from Amsterdam, we (slightly) entered the Bermuda Triangle and survived!

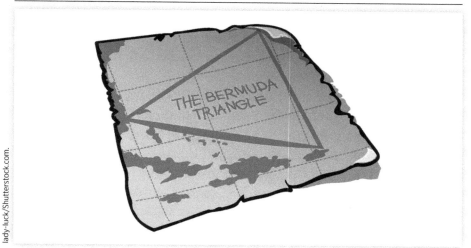

lady-luck/Shutterstock.com.

Project Team

Forming a team is not an easy task. My organization was a normal organization, and I wanted to run the project. Rule number one: A normal organization always has problems with project organization. Rule number two: Project organization always has problems with the normal organization. I had to work in that environment. It was not my first time, and I knew my organization would be displeased with me very soon.

The project began with the engineering specification. Not many engineers were in a management position within engineering, and not many engineers were in management positions in the maintenance department. Most managers had studied business science or politics, and some were economists. This led to communication problems. In this environment, the engineering manager was unwilling to assign engineers to projects for a third-party customer. The maintenance department management was unwilling to perform tasks other than standard D checks. The sales department wanted a D check, and modifications were a necessary evil. D checks brought money; modifications were too complicated (and not as lucrative). Engineers and hangar technicians (techs) wanted to run modifications because ordinary D checks were boring. My job was to manage all parties and motivate them to get the job done. As a project manager, it's difficult to make many friends. I was aware that everyone on the team would hate me and love me at some time during the project.

To get the machinery rolling, I had to go to upper management and escalate the issue. The first step was to go to the CEO together with sales to get the engineering managers in motion. They would get the engineers to begin working on the specification. Based on the specification, I can calculate the modification costs and submit the proposal to sales. Sales adds 15% and sends the proposal to Mac, the CEO. Internally, there was a lot of back and forth between the many individuals involved. The complaints were always the same: too many engineering hours, too many project management hours, not enough hangar hours, expensive materials, and so on. Finally, the glorious day arrived when the proposal was sent to Mac. To my surprise, Mac returned the signed proposal within an hour. It was about 1M USD, and the contract was approved within minutes and without negotiation. Mac was tough but he was also quick to make a decision.

Focus Modifications

Now the actual work could begin. The engineers would do their jobs, but I had to hold all the bits and pieces together. After many hours of discussions and coordination with the engineers, the list of materials was created (for me), the list of subcontracts was created (for me), and the certification plan was created (for me). The next step was to get all the subcontracts done and to get all materials and drawings ordered.

The first aircraft arrived in Amsterdam from Roswell, and it was parked in front of my hangar. This made life easier. Engineers could now visit the aircraft and confirm whether the situation in the aircraft matched the drawings. We could also make sure that the components installed were of the same p/n and modification status as in the IPC.

Meanwhile, my sales buddy Bastian told me that Mac was going to buy two more 747s from Malaysia Airlines. I decided to send Mac an urgent message.

Dear Mac, the 747s are delivered in two different cockpit configurations with respect to the switch convention. There is a so-called "Forward on" configuration and "Forward off" configuration. Meaning that all cockpit switches are either forward on or forward off. The FAA does not allow pilots to fly both configurations. Some people call it "Switch Convention." I mean a pilot cannot fly one day in a 747 with forward on switches and the next day with forward off. That brings confusion, and it is dangerous. Please make sure when you are buying the next two aircraft that they will be the same as the first two. The first two are in "Forward on" configuration.

Within an hour, Mac replied "MJ, I already signed the contract for those two aircraft. We will check the configuration." The next day, Mac reported that the two Malaysian aircraft were "Forward off." Aargh! This was one more worry. It was not just a matter of adding the new effectivity, but all control panels in the cockpit would have to be replaced or modified in these two aircraft. We would have another expensive proposal for Mac to approve.

In the meantime, the D check started on the first aircraft. The modifications were also in the works, although some material had not been delivered. This is part of my job as the project manager.

The aircraft is in the hangar, the power is off, and a lot of work is going on. During the first period of the check, I could only sit in the office and call different parties to push them to send the material. We often had problems with the rotables (LRUs). Even if they arrived on schedule, there was a complicated logistics chain that introduced delays. Some people said that there was a wormhole at the airport in Amsterdam. Theoretically, a wormhole might connect extremely long distances such as a billion light-years or short distances such as a few meters or different points in time or even different universes. The boxes of material we needed for our modifications disappeared at the airport and then reappeared someplace nobody expected.

After four weeks, the modifications were done, the D check was nearly finished, and the power was connected to the aircraft. This was the moment the testing started (EGPWS, ACARS, GPS, ELT, TCAS, ELS). Although the modifications were based on existing STCs, it was still necessary to do a thorough test to make sure that everything would operate smoothly for the next five or six years. The first aircraft modification was complete and that craft was parked in a different hangar. The second aircraft was not coming to Amsterdam for the D check. Mac decided to ferry it from Roswell to Florida and perform the D check and all modifications there. My management didn't like this at all, and it also became another concern for me. Now I had to make sure that all materials were sent to Miami (another possible wormhole). Mac also requested a few techs for on-site support. I had to make staffing choices, because too many techs volunteered to go to Miami.

A fight with my own managers soon followed. Mac was paying techs and providing hotels and transportation, but my management still didn't want to send more than two people to Florida. This was a stupid decision, and it was very disappointing for the techs from the hangar who wanted to travel to Florida for that on-site experience. After another escalation with some local managers, two people per shift were assigned to go. Not me, of course—I had to support the team in Miami from Amsterdam. My friend Laurine was there, which was reassuring.

The second aircraft was undergoing modifications, and my on-site team was enjoying Miami. It was a great experience for them. I also learned something new on this project. Laurine told me that there was a problem with the ELT modification. ELT is a simple modification. I had decided that the modification could be approved per FAA Form 337. That would be enough, because the regulation says that Form 337 can be used as a one-time approval, and it is acceptable for the administrator. But Laurine told me that the administrator should be given advance notice of the approval plan. I didn't know that the administrator had to agree in advance to accept Form 337. I was lucky to have Laurine there in Miami; he solved the problem for me.

The third and fourth aircraft were also not modified in Amsterdam. Mac decided to send them to Kuala Lumpur for the project, but he wanted the support of one engineer and six techs. One of my engineers went to Malaysia for 12 weeks to work on this project. By then, it was proven technology. All the modifications were correct, and even the additional modification on the switch convention was successful. Every aircraft was different, but we managed to modify each one and get it airworthy. Or did we?

The first aircraft was still parked in the hangar in Amsterdam. The D check was done, modifications were complete, but the aircraft remained parked. The tanks were leaking, and fuel was dripping from the wings. A blame game began between Mac and my management, though I was unaware of it. After a discussion with the Focus air Cargo representative, we went to the hangar and inspected the aircraft. As a result of that inspection, we found 14 problems with the aircraft. It was ridiculous that previously, no one had compiled such a list. I sent an irate message to my upper management. The subject of the email was: "The airplane will not depart unless," and then I provided the list of 14 problems. It wasn't rocket science; the issues were quite simple. Just to mention a few: Hatches were not painted with the text indicating what was behind them, the calibration date of the air data test set was needed, there was a leak in the fuel tanks, and so on. I enlisted the aid of six techs, and we fixed every problem within two days. Go bird, fly away! (See Figure 12.3).

FIGURE 12.3 | Focus Air 747 Freighter, finished, polished, and ready to go.

Courtesy of Piet Alberts.

This was the most difficult project I ever worked on. Not because of the technical aspects of the work, but rather, there were too many internal political issues to manage. It was a very unhealthy situation. My project management skills were exercised to the limits. Still, I managed to *focus* on the Focus project. All four 747s were made compliant with the latest regulation and airworthy. All modifications were FAA-approved, and Mac was happy. Unfortunately, my own management was not happy with me because I was perceived as favoring the customer. As I indicated earlier, people from normal organizations don't like projects, and they *really* don't like project managers. Shortly after the project's successful completion, my manager told me: "All modifications are now done. We don't need you anymore. Tell me what you would like to do, and I will transfer you to another job." My manager was a politician, and I was an engineer. We were not a compatible pair.

End of Focus

You might think that the story about Focus and "my friend" Mac has come to an end. But not at all. Years later, in the fall of 2015, I got a call from Mac. He told me, "I will come over to The Netherlands to talk to you. I need your help." A few days later, he came to Amsterdam from California to talk to me. Here follows the final chapter of the Focus story.

Mac sold the 747s and ceased operations with Focus. He was then CEO of a company building electronic devices used to measure various parameters at oil wells. They install the devices on oil wells. The device records measurements, receives time and position information from the GPS, and sends the package with that information via the Iridium satellite to the ground station. After analyses of the information, the results were stored in the cloud.

And now the reason for Mac's visit: The 777 of Malaysia Airlines Flight 370, was a scheduled international passenger flight that disappeared on March 8, 2014. While flying from Kuala Lumpur International Airport in Malaysia to its planned destination Beijing Capital International Airport, the aircraft was lost and never found again.

Mac's idea was to build a device with his technology in the shape of an aircraft window that could be installed in the aircraft instead of one of the passenger windows. The device would consist of two antennas and a processor unit. One of the antennas would receive the GPS position and time signals. The position would then be stored for 5 minutes in the unit's memory. Every 5 minutes, the information would be sent via the Iridium system to the cloud. If an aircraft was lost, you would be able to pull the data from the cloud and track it down to the moment of impact. The unit is very inexpensive, easy to install, and works independently. What a great idea! I impressed Mac enough in 2004 and 2005 with his difficult 747 fleet project that he felt I was a good resource in 2015. We started to collaborate, but unfortunately, there was no interest in such a system. Everyone in the industry was looking to Boeing and Airbus and waiting for their solution and for the FAA or EASA mandate before taking action.

We are still waiting. Malaysia Flight 370 is nearly forgotten. Sometimes it seems like no one values the pioneer. We could have offered a solution to this problem, but the industry missed the opportunity. Perhaps there was not enough *focus*.

13

Fight on the Flight

Not long after the 9/11 catastrophe, the authorities (FAA and EASA) issued the Airworthiness Directive requiring the installation of enforced cockpit doors. After the cockpit doors mandate, a request was issued by the authorities to install cameras in the cabin.

At that time, I was sharing an office with an engineer named Pat. We had a lot of modification work for the wide-body fleet. Some modifications were in the package for our fleet. Usually, when the aircraft was due for a C or D check, modifications were inserted into the C or D check package. C and D checks are based on the airframer's Maintenance Planning Document (MPD), which is a document provided by the airframer upon purchase of the aircraft. Based on the MPD, engineers within the airline design the C or D check package. Then they add some items based on experience or the airframer's service letters to complete a customized package for a C or D check.

On a separate note, engineers review the SBs and propose modifications. Sometimes they issue the proposal for an LRU modification and sometimes for an aircraft modification. These proposals usually land on our desk (Pat and me). We then make sure that the price is calculated (time and materials), and then we take the finished proposal to the proposal board. The proposal board was a completely unnecessary department full of commercial types who had no understanding of the magnitude of modification work. They only cared about reaching their goal and generating turnover. They take our modification proposals to the fleet, discuss them, and try to sell them. That was the ball game between departments within the company's business units. The technical department was the business center and was supposed to sell modifications to the fleet owner which a was different business unit.

Many times, my sales managers gave away the modifications for free, so I called them "giveaway managers." They didn't like the nickname, and we had the basis for a bad relationship. The efforts of my engineers and techs were just given away as a gift to fleet owners, who bragged later about how they squeezed us engineers and techs. I obviously didn't have much sympathy for the fleet owner's department (our internal customer).

Between implementing simple modifications, we sometimes got a bigger project. Actually, major fleet modifications were a regular occurrence. An example of a large modification project was the requirement for reinforced cockpit doors. Shortly after that came the request for cabin surveillance. We agreed to flip the coin to see who would manage the larger project and who would manage the smaller, less interesting modifications. We flipped the coin for the reinforced door, and I lost. Then we flipped the coin for the cabin surveillance, and I lost again. Pat was left to wrestle with those two modifications—especially with the cabin surveillance. The reinforced door was easier: You buy the airframer's master change (what we used to call Master Charge), and you get the FAA- and EASA-approved SB and all the required materials. Piece of cake.

The cabin surveillance modification was a full-blown project. It included the specification, vendor selection, contract negotiation, and, finally, installation in the aircraft. Pat was in my office, so I was able to follow all the discussions and concerns of the project manager.

I believed that I would be finishing the project. That was my destiny. I had been in the same situation on several previous projects. When things start to get complicated, the project manager moves on to another job. They did all the fun stuff, like traveling to vendors, vendor selection, negotiation, and so on, and when the project became too technical or frustrating, they suddenly moved to another position and … guess what? I was given those difficult projects to finish. I usually didn't mind, because it was a great learning experience for me.

Transaero

When Pat was already one year into the project, I received an email from my Russian friend Vadim Demin, whom I called VD. (I will use VD in the rest of the chapter.) VD had a great idea: He and a friend had established a company Amstel - M, and they were looking for a technical director to manage multiple projects for their fleet. The name of the company they were actually working for was Transaero. Transaero went bankrupt six or seven years later, but, when VD contacted me, they were quite successful, and they were flying all Boeing aircraft. The fleet count was about 22 aircraft. They had a problem, however: They didn't have many contacts in the US and EU to work with. This was the reason they wanted to have me on their team.

Around this time, I was in Jakarta, Indonesia, with my friend Danny from sales. We were staying at the Hotel Shangri La. After a few beers in Bats (a nightclub), we walked to our rooms, and along the way, I told him about VD and the opportunity at Transaero. He advised me to send a letter to the KLM human resources (HR) department and request permission to do consultancy work in aviation. This was the best way to play it clean. Danny also has his own company, importing all kinds of supplies for hairdressers in The Netherlands, as well as importing sunglasses. Many sunglass companies don't like to work directly with those small sunglass merchants who sell sunglasses in the farmer's market or small kiosks. Danny filled that gap by supplying these vendors. When I returned to The Netherlands, I immediately sent a letter to HR asking for permission to do consultancy work. A few days later, HR responded. They said OK, but I needed to respect my work and rest time schedule. Of course, I understood this. My employer would not like it if I worked the whole night and slept in the office. Later, I heard that the company had a bad experience with staff who were working all night in Amsterdam nightclubs and reporting to work in the workshop in the morning with no rest time. That is not the lifestyle for me.

With HR's approval, I said yes to VD, and we were in business. Working with Russians is different from what I was used to. There was no contract, just a verbal agreement. I understood this, because I had worked like that in Croatia. When working with Croatians or Albanians, a verbal agreement is enough. In the case of VD, I didn't have much to lose. They didn't ask me to invest my money in the company. I took VD at his word: They would pay me, I was certain. I had established my Russian connection.

Why did Transaero need me? VD's company was flying from Moscow and a few other cities to the Mediterranean and North Africa. These were chartered flights with thousands of tourists going on a two-week holiday to cities along the Mediterranean or the Red Sea: Tourists seeking warm weather, happy Russians celebrating their holidays. "Celebrated" might be the correct tense, because the celebration started on the airplane. The passengers drank a *lot* of vodka, which they purchased from the crew or smuggled onto the plane. It was not unusual for passengers to start to argue and sometimes physically fight. It was totally out of control. The operator's fleet was growing because of good business, but those in-flight fights were problematic. According to VD, the company's lawyers were in court every week arguing about who started the fight, who delivered the first punch, and who would pay the damages.

My first suggestion to VD was, don't serve them so much vodka. But the company management didn't want to hear that. They were afraid that passengers would go to another charter operator to enjoy a booze-filled flight. This was typical Russian company management. They said that the top of the company was a three-headed dragon: One head was the owner, the second head was his wife, and the third head was the managing director who did all the work. The owner and his wife were there to get huge salaries, and the managing director was running the business.

We started a cabin surveillance project for Transaero. This was a completely different project from the cabin surveillance project managed by Pat. The Russians didn't have to comply with the EU requirements governing the installation cabin surveillance. The justification for this project was to record the fights among drunk passengers. They wanted a server that could record video for at least 20 hours, two cameras in the cabin, and night vision to monitor passengers in the dark.

This was the project directive: three requirements with no further explanation. They wanted the video to record all the time. After 20 hours, it would start a new loop. If the crew reported a fight in the cabin, the engineer would intercept the aircraft at the gate after landing, connect a laptop to the server, and download the video. This was a very simple project. I made a few pages of specifications and descriptions of the system and sent these to the Russians. The cockpit crew was not happy with the specification and started to complain, because they wanted control over the surveillance system. They wanted the video screen and control panel in the cockpit to have the capability to switch between the cameras and to dim the display. This was possible. Additional money was required, but the Russians were able to pay, so the change was fine by me.

After some consideration, I decided to contact a UK company that I knew was an expert in the surveillance business and reputable as well. My colleague from the office, Pat, decided to contract a different vendor for his project to save money. Later, he realized that more work was required of him to complete the project with that vendor, but by then, it was too late. The UK company was not new to the surveillance business. They were selling surveillance systems for banks and offices. Their biggest contract to date was with the City of London when they installed 20,000 surveillance cameras in the streets of London for the 2012 Olympic games. At that time, they had just started in aviation and were becoming successful very quickly.

After some discussion with the UK company, I put together a proposal for cabin surveillance for my Russian friends. VD told me that the proposal should go in the name of Amstel-M

Company, which was Russian-owned and based in The Netherlands. One of his friends was the owner. The proposal went to Moscow, they approved it, and after that, they issued one purchase order per aircraft. There were four modifications applicable, one for each aircraft type (737, 747, 767, 777). First, I was the purchasing manager and then the proposal manager. Now, I became the modification manager, and later I would also be the support engineer.

Doing such projects was easy for me. Every project has several phases. When an engineer is doing their job properly, they will be able to recognize the shift from one phase to another. In this case, it was easy to see that after the specification phase and financial approval, the design phase began. I had to answer many questions asked by the UK company (manufacturer) and many questions asked by the Russians. Both listened and followed my instructions. I truly believe that the success of a project is 80% dependent on the project manager. The success rate was good on this project, because I had them all eating from my hand.

After six months, the first kit was ready. The first installation on the 747 was planned in Moscow. I asked VD, "Are you sure?" "Yes, I am sure," was the answer. The kit was shipped to Moscow but never arrived at their facility. It got stuck at customs. It was confiscated with the explanation that the package contained spy equipment. Of course, cameras, a display, a control panel and a server and some wires—these are spy equipment. Transaero intended to spy on their passengers.

We learned a valuable lesson: Don't send surveillance equipment to Russia. It took VD six months to negotiate the release of the first kit. In the meantime, the modification was running successfully at other Maintenance Repair Organization (MRO) facilities. Once the system was certified for each type of aircraft—we ensured the design was correct and the drawings were without mistakes—the rest was easy. VD just had to provide the address of the MRO and tail number, and I made the arrangements for the kit to be sent there. We completed the modifications in Paris, Riga, Belgrade, Singapore, Kuala Lumpur, Miami, and somewhere in China (Figure 13.1).

FIGURE 13.1 Transaero 747 with my cabin surveillance modification.

NYC Russ/Shutterstock.com.

Windfall

I had a lot of fun on the project and didn't even ask for payment. At a certain point, VD asked me for my bank account number. It was near Christmas, and I thought it would be good to have some spending money. The AM Company transferred the money on December 31st, and I received it on January 1st. It was a large payment. I worked months for it, but it still didn't feel quite right. I didn't have a contract, they never asked me for an invoice, and they paid me a full year's salary in one shot. This is how it works in Russia. My problem was that I had to pay 43% tax in The Netherlands on that large payment. But the Russians were smart about it. The money was in the cloud and, at the end of the year, was not in their account. I received the payment in January, though it was payment for the previous year. I knew that I could now take some of my expenses and subtract them from the total sum to make that 43% tax payment as small as possible. The costs had to be the expenses for running the business—business trips, dinners, car expenses, attending conferences, computer, and various peripherals. If it was business related, it was a deductible expense. It was good for me to get paid in January and have the whole year to generate expenses.

Now the window of opportunity was open for me, and the fun started. The first thing I did with that extra income was training: Effective negotiation. I saw it in an advertisement on a US domestic flight. The instructor's name was Chester. He was one of Ronald Reagan's negotiators with the Russians to limit nuclear weapons. He was very smart and had designed a great course. Conveniently, I could take the course right here in Amsterdam. It was a three-day course, and I happily paid 2,000 USD for it. I have no regrets about the cost, because, with the tips and tricks about negotiations and the practical exercises, I became skillful at negotiations. It was the best investment I ever made.

I also took a trip to Dubai for the Honeywell conference. I had the opportunity to see Dubai and even took my better half with me. After that, there was a Boeing conference about aircraft modifications in Istanbul, which was just one day after the Parts Manufacturer Approval (PMA) community conference. I had two conferences in one trip and saw Istanbul as well.

Le Mans VIP

The Transaero fleet had grown to 60 aircraft, and we were buying more and more kits from the UK. The company called me with a surprise: The UK company owner was a fanatic car racer who owned a team and a car brand name: MG Lola. He was racing the Le Mans series and then the famous 24 hours of Le Mans. They offered me a four-day trip to Hungary to the Grand Prix circuit Hungaroring. I would attend as a VIP and have access to their VIP booth. I was not allowed to accept such gifts when I was representing my airline, but this was different. My airline had no business relationship with the UK company, and I didn't have a contract with anyone. So why not? Let's see that car and Hungaroring! In a couple of days, I received my e-ticket and hotel reservation in

Budapest. Hungaroring, here I come! They told me that VD would also be there. Finally, I would meet VD, my Russian connection.

It was just a 1-hour 30-minute flight to Budapest. A car was waiting for me at the airport. It was the second time I had a driver show signage with my name on it. I arrived at the hotel and met a few people from the UK and, of course, VD, my Russian connection. A large group of the UK company's customers were also there—a lot of non-aerospace people. Only VD and I were in aerospace (Figure 13.2). They got all of us together, brought us to a meeting room, and locked the door. Then they started to talk about their products. The woman who was presenting gave a long, boring talk, but they were paying for the trip, so we had to be polite. VD and I were exchanging text messages like, "Boring! OMG, if she keeps on talking, I will die."

FIGURE 13.2   Enjoying my VIP status at 24 hours Le Mans.

Frolphy/Shutterstock.com.

Over the next few days, we enjoyed beer and car racing (Figure 13.3). Each team at the Le Mans series has three drivers. Our team had the boss (he got to drive because he was paying for it all), a

FIGURE 13.3 Engineers always have fun (VIP at Hungaroring during Le Mans series race).

Courtesy of Marijan Jozic.

Brazilian guy called Tommy, and Ben Collins, who is "The Stig" on the TV show *Top Gear*. This was a spin-off from the aircraft modifications project. Are we having fun or not?

Suitcase Surveillance

After those days full of fun and fast cars, we had to go home and continue the project. The Russians were buying more and more aircraft, and my project was growing bigger and bigger. I had also started to do other small modifications to reconfigure the aircraft. Every time they leased a new aircraft, they had to change the tail number. The tail number then got the new 24-bit ICAO code, which had to be preprogrammed in Mode S transponders, SATCOM, and ILTs. This was a small modification and easy money. We were running smoothly as a team. Money was flowing, the customer was happy, and the fleet was flying.

At one point, VD came up with an addition to the cabin surveillance. In some airports, the people who loaded suitcases in the cargo compartment were opening and stealing from the suitcases and bags. The Russians wanted spy cams in the forward and aft cargo bays of their aircraft. This was possible because we had four more channels available on the server. With just four cables and four small cameras, it was doable.

When I was wrapping up the additional proposal and starting preparations for a new project, the political situation changed. An embargo was imposed on Russians for invading the Crimean Peninsula. VD's company was banned from flying to many Mediterranean airports and the Red Sea area. With no flights, there is no cash flow. A few months later, VD's company was bankrupt. That was the end of my cabin surveillance project after a great four years of moonlighting.

14

Conferences

In the early 1990s, I attended a few conferences. Two were related to RVSM, one was related to SATCOM, and one was related to EGPWC. The RVSM conferences in Copenhagen and St. Petersburg were productive, and I learned what was needed for installation and activation of the new system.

The SATCOM conference in London was organized by Inmarsat and was nothing special. Many Chief Operating Officers (COOs) and high-ranked individuals used the conference to brag about how good they were and how important their company was.

I often think of IBM president Thomas Watson, who said in 1943, "I think there is a world market for maybe five computers." Or Ken Olsen, founder of Digital Equipment Corp, in 1977: "There is no reason anyone would want a computer in their home." Even Bill Gates said in 2004: "Two years from now, spam will be solved." Perhaps the funniest was Alex Lewyt, president of Lewyt Vacuum Cleaner Company, who said in 1955: "Nuclear-powered vacuum cleaners will probably be a reality within ten years." Well, it is now nearly 70 years later, and I have not seen or heard of such a vacuum cleaner (thankfully).

At the conference in London, head honchos provided predictions about technology development. Inmarsat's president complained that he could not send an email during his flight from Singapore to London. That, for him, was a big deal—8 hours flying and not being able to send an email! In all the years after that event in 1990, I never heard that passengers had a big desire to send emails in flight. Business class passengers are generally happy to be able to sleep undisturbed for a few hours. Many flight attendants told me that the easiest job is to be a flight attendant in business class, because the majority of business people first eat and, after that, go to sleep and ask not to be disturbed until the landing.

The president of Iridium announced in his speech at the conference that, within five years, there would be 700 million Iridium phones on the planet. Connectivity is essential, and Iridium can provide it. Five years later, they had sold only 2,000 phones and were nearly out of business. Both companies bet on satellites and underestimated the terrestrial network.

This was also the reason Inmarsat and the service providers kept prices so high. At that time, the first SATCOM minute was 15 USD, and every next minute was 10 USD. That is just too expensive for the average passenger, and if business people only want to sleep during the flight, where is the market?

The third conference I attended was the EGPWS conference in Palma de Mallorca. This was a productive conference, because suppliers from Boeing and Airbus explained everything about the system and aircraft installation. I also had the opportunity to meet all the important players needed for the installation of the system. In addition, there was a test flight organized by AlliedSignal. During the conference, groups of five people were allowed onto the small business jet, which was flown by a very tall guy who called himself Little John. Little John took us from Palma de Mallorca Airport to the western side of the island to fly alongside the mountains. We could look through the cockpit window as we approached the mountain. After the call out, "Terrain, Terrain!," he made a right turn and came back to approach the mountain and do the left turn. I got a pretty good idea of the importance of EGPWS on the distance of the mountain before impact.

I also wanted to go to the AMC (Avionics Maintenance Conference), but my requests to attend were rejected by my managers every time I asked. I had heard a lot about this conference. I studied reports from the conferences and commented on the working papers, but every year, somebody else was sent, usually one of my fellow engineers who had been in engineering for at least 15 years. So due to seniority, they rejected my applications to attend. At one point, a couple of managers went to the conference without even knowing what AMC was. They were not very technical and went to the conference unprepared. They came back telling everyone that the conference was a waste of time! If you said that to Dutch management, they'd immediately scrap the conference from the travel list and brag about how they saved money on the engineering department's business trips. I seemed always to be in the wrong place at the wrong time.

After 10 years of trying, my request was awarded in 1999, and I was allowed to go to the AMC in Baltimore. I had a nonstop flight to Baltimore, and I got a standby ticket for a flight to Baltimore and my first AMC. I had done quite a bit of preparation for open forum discussions, so when the discussions started, I was ready. That evening, I met a guy from United Airlines who told me, "Hi, man, you had quite a lot of microphone time during the open forum." I was surprised and pleased that I had been noticed. Later on, the AMC executive secretary said to me, "Welcome back, flying Dutchman!" He was happy that after a few years' absence, my airline had started to attend the AMCs again.

It was a great feeling to be among engineers. I felt immediately that I was in my habitat. The AMC was the mother of all conferences and the most important event of the year. The next year, it was held in San Francisco, hosted by United Airlines. I already knew a lot of people and had the opportunity to visit the United Airlines facility. What an experience! In 2001, I attended the first AMC outside the US, in Hamburg, Germany. The conference was hosted by Lufthansa, and it was a historic moment.

Plane Talk

At the Hamburg AMC, something remarkable happened. I was sitting on the couch in the Smiths Industries hospitality suite eating ice cream with the *Plane Talk* editor-in-chief and executive secretary of AMC. We were chitchatting, and I told him that I would like to write an article for *Plane Talk*, the magazine of the AMC. I was afraid that I would make a big mess of it, but there was that little voice in me saying, "Why not ask?" He said, "Go ahead, write something and just send it to me. I will check it and tell you my opinion."

That was my plan, but I forgot about it for some time. Three months later, I was in Croatia on holiday and had my laptop with me. My holiday was three-and-half weeks, and after the first week, I was bored. I thought about my plan with the editor of *Plane Talk* and started to write. My first article was titled, "Is It Fun to Be an Engineer?" It was a six-page article, and the only way to get it to the editor was to save it on a floppy disk, go to an Internet café, and send it via email. We used email back then, but Wi-Fi was unusual. Mobile phones were quite common, but smartphones were not, so the Internet café was my resource for Wi-Fi. Approximately one hour later, I got a call from a very excited editor. Yes! He wanted to publish my article in the next *Plane Talk*. This is how I became a writer.

An old wound of mine was healed that day. Twenty-five years earlier, I was in Croatia (at that time, Yugoslavia), a 16-year-old boy attending college. My professor didn't like my straightforward writing style. He wanted stories with a lot of adjectives and other bells and whistles. Some students in my class were good at that style, but I couldn't even understand their sentences. This professor's criticism killed my writing for at least 25 years. He was truly a terrible professor, but as a young man, I was too polite to tell him that (and now it's too late, because he is no longer alive). I will always be grateful to my *Plane Talk* editor, who helped me rediscover my writing skills and dreams and published my first article in the summer of 2001. About a week after *Plane Talk* was published, I got an email from Honeywell. They had a company magazine, *Data Bus*, and the editor of *Data Bus* asked me if I would grant him permission to republish my *Plane Talk* article. I did. Soon, the editor of *Aviation Maintenance* sent me an email too. He asked if I would like to write articles for him, and I agreed. After that, I had to submit one article per month to *Aviation Maintenance* and six articles per year to *Plane Talk*. I had a writing career.

This is the power of the AMC: If I was not in Hamburg eating ice cream with the editor of *Plane Talk*, I doubt my writing career would have happened. I was writing articles for fun and doing my engineering and project management at my airline, and, of course, attending conferences.

The *Plane Talk* editor died unexpectedly and far too young. He was the driving force behind the AMC, and everyone felt like they knew him well. He also seemed to know every individual at the conference. His death was a great loss for the industry. To honor him and his boundless passion for aviation, the Roger S. Goldberg Award was established by the ARINC Industry Activity (the organizer of the AMC conference) for individuals in avionics with a demonstrated passion for aviation. The award has been presented each year since 2007.

The next important AMC for me was in Montreal. It was held in a Sheraton downtown, and upon entering the ballroom, one of my friends said to me, "How do you feel? Today is your day." What was that supposed to mean? An hour later, they called my name, and I had to go to the podium to receive the Volare Award (Figure 14.1). I had always looked up to the winners of the Volare Award with great respect. Each year the avionics community selects three individuals who have done extraordinary work to receive the Volare Award. There is a Volare Award for engineers in the supplier's category and engineers in the operators (airline) category, and there is a Volare Pioneer Award. I was surprised to find myself among the extraordinary individuals who have won this award. And it was only my fifth AMC! I was 48 years old, and somebody commented, "You are too young to get a Volare Award." I thought I was at the top of my career.

FIGURE 14.1   Volare Award recipient in 2004.

Courtesy of Marijan Jozic.

Courtesy of Marijan Jozic.

A month later, it was announced that I was nominated for the Aerospace Journalist of the Year Award, which was presented on the first day of the Farnborough Airshow. I went to the ceremony at my own expense. I was one of five nominees, and I felt like I was already a winner. I was not even a journalist, but an avionics engineer and a project manager. I didn't receive that award, but had a wonderful time just being there. The following year (2005) at the Paris Air Show, I was nominated again for one of my articles from *Plane Talk*. I did not receive the award, but it was great to attend the ceremony and to be nominated.

The same year (2005) at the AMC in Paris, I applied for a position on the steering committee of the AMC. The engineers voted, and I got a seat. Now I could actively contribute to making the conference even better and more exciting. A few conferences later, I even became the Vice Chairman of the conference. This was a great honor and responsibility. We had to keep the conference exciting and productive for attendees and that was very challenging for me.

At the AMC/AEEC conference in Anchorage, Alaska, in May 2012, I decided to run for chairman. The engineers voted and made me the chairman. Around that time, I started to keep track of the savings each conference brought to my company, KLM. After five years, I found that at each conference we gained 250,000 USD. At the conference, I learned about cost avoidance, new repair methods, and cost savings by addressing AMC questions in the open forum. Conferences helped KLM make significant savings.

With this information in mind, my management no longer complained when I submitted a request to travel to a conference, to steering committee meetings, or to working groups. Along with being chairman of the AMC, I was also chairman of a few ARINC committees producing standards. Sometimes, I still ask myself: Is it fun to be an engineer? Believe me, this engineer had a lot of fun doing projects at KLM, managing conferences, writing articles, gaining a lot of knowledge, and creating a large network.

Everything was lined up for a meaningful conference in Milwaukee in 2017. They told me there would be no awards, but on the first day, they put me on the spot. I received the Roger S. Goldberg Award (Figure 14.2). This had to be the top of my career— it couldn't get any better! A beautiful shiny Roger S. Goldberg Award for boundless passion in aviation. This award was extra special to me because it was awarded by my peer engineers, and they only give an award if they are convinced that the person deserves it.

FIGURE 14.2   Roger Goldberg (RGB) Award, presented by Dean Conner.

Courtesy of Marijan Jozic.

Courtesy of Marijan Jozic.

MMC

In 2017, a new ARINC Industry Activities conference, the Mechanical Maintenance Conference (MMC), started. This conference has an interesting backstory. A few years prior to the first conference, avionics engineers reported that their team members who worked on mechanical systems were upset because they didn't have a conference like the AMC. Air France had asked twice, once at two different AMCs, about the possibility of a conference dedicated to mechanical systems. We discussed this in the AMC open forum and then later at the AMC Steering Group meeting. We asked ourselves if and why a mechanical conference was a good idea. We decided to add one day to our AMC Steering Group meeting in October 2015 to thoroughly discuss what we named the MMC.

We knew that starting the conference would be difficult, but we took up the challenge. The first step was to organize an exploratory meeting and invite professionals in engineering and maintenance of aircraft mechanical systems. This was no small undertaking, given the potential impact on the industry. The biggest challenge was attracting enough airlines and their respective engineering representatives to participate in the conference. We knew from our experiences at the AMC that if airlines attend, the suppliers will attend too. The MMC exploratory meeting took place in Seal Beach, California, and we invited a few airlines, airframers, and equipment suppliers to attend. After the MMC Steering Group meeting, we met with suppliers and airframers. Airbus announced that they would support such a conference, and after just a few hours of meeting, we felt that we were on the right path. The decision was made—a new conference would be organized by ARINC Industry Activities. We planned to systematically map the challenges we faced in starting the new conference, and we worked through them, just as the AMC planning committee does every year. We were now past the point of no return; the conference was on the calendar, and there was no turning back. Every year, AMC saves time and money for the airlines by finding solutions to avionics issues. We believed that the MMC conference could do the same. The key is cooperation between airlines, airframers, and equipment suppliers.

So we did it—we launched the MMC. The first was held in Cleveland and was a big success. The majority of attendees were mechanical engineers, but the conference was moderated by us avionics engineers. By planning and starting up this new conference, we showed that where there is a will, there is a way. Shortly after the conference, at our winter meeting, the steering committee decided to integrate the MMC into the AMC and open it for all engineers and all ATA chapters (Air Transport Association 100 System Codes). The only area not included is engines and primary structures of the aircraft. We also decided to change the name.

Instead of Avionics Maintenance Conference, the new name would be Aviation Maintenance Conference. This was a timely decision, as it integrated the old with the new.

We had a full-blown Aviation Maintenance Conference in Dallas in 2018. I was in my seventh year as a chairman. The AMC was nearing its 70th year, and I was the only person who had been chairman for so long a term. The nearest contender was Walter D. Rollick from Piedmont Airlines, who was chairman from 1950 to 1955, so I was the holder of the record. I felt that my time leading the conference should come to an end, and I announced in Dallas that the next AMC would be my last as chair. Everyone was flabbergasted, but my rationale was that it is best to leave at the top of my career, and it is best that I decide when to leave. I've been told that it is best to stop when it causes a little heartache. That day in Dallas was the right time for the announcement, and the 2019 conference in Prague was my last as conference chairman.

AMC Prague

I wanted to leave in full glory, and my wish was that KLM would host the Prague conference. But the KLM COO didn't want to finance the conference unless it was in Amsterdam. He told me this every year, so it was not a surprise. I disagreed with his decision for a number of practical reasons. In the first place, the AMC concept was to have a conference ballroom and hotel in the same building, which made it easier to organize the hospitality suites and the showcase reception. There is no hotel in Amsterdam with the capacity to support this. Even large hotels didn't want to provide more than 200 rooms, and our usual need was 550 rooms. Going outside Amsterdam for a hotel caused transportation complications. The second problem with Amsterdam was the high room rate. All hotels in Amsterdam were above 200 USD per night, and they wanted at least 30K USD in advance to hold the rooms. I realized then that the Dutch business mentality is poor. About 300 years ago, the Dutch were known to be practical and business savvy. In my experience, there is not much of that business acumen left. The AMC has a solid track record. Hotels in the US called us every year and asked us to bring the conference there. They know that we spend a lot of money on rooms, food and beverages, and hospitality suites. About 650–750 engineers attend the conference each year. This is a great number, but apparently, it's not large enough for Dutch hotel owners. I could not do much more than lobby internally to convince my management to host the conference in Prague in 2019.

I was reading a book *Start with Why* by Simon Sinek in which he explained that it is important to start the communication with *why*. I thought: Let's give it a try and send the COO a letter. I was sick of talking and begging him and his VPs to host the conference. I thought, Let's send a memo and start with WHY! It was just a one-page memo that started with *why*:

> Whatever we do at the AMC, we are always challenging the status quo. We always ask ourselves if we can do things better and more efficiently and cost-effectively.

His short message came back: Go for it!
And I got myself a conference.

I had six months to make all arrangements, as the host of the conference must do. The biggest challenge was to book a keynote speaker, because all my higher managers were suddenly engaged in other things. I thought it was sad that my airline was the conference host and someone outside the company was the keynote speaker. But it is what it is. At the conference there was one more surprise for me. The steering committee has the duty of selecting an aerospace man of the decade, and they unanimously selected me (Figure 14.3).

FIGURE 14.3 Man of the Decade Award.

Courtesy of Marijan Jozic.

It was a great honor and the crowning achievement of my work for the aviation industry. I assumed again that I was now at the top of my career. The Prague conference was a great success for me and the AMC and ARINC Industry Activities. After the AMC conference in Prague, I handed the torch over to my successor from American Airlines.

FOM

One interesting spin-off to my work in the aviation industry and in attending conferences was how I became involved in obsolescence. There was an obsolescence conference in Amsterdam called Future Obsolescence Management (FOM). Nobody from our purchasing department wanted to go, so they gave me the ticket. It was a one-day event at a museum in Amsterdam. There were six speakers, and the audience was allowed to ask questions. I asked a few questions and was engaged in a discussion. At the end of the conference, a gentleman came up to me and asked if I could speak at their next conference, which would also be in Amsterdam. I said yes. We were working on the ARINC obsolescence standard at the time, which would be quite useful to the presentation. What could I possibly lose? I had one year to prepare.

I was working hard to finish that ARINC standard and thinking about my presentation. About two months before the conference, the conference organizer offered me a professional instructor named Nick Graham to teach me how to present. I took a few lessons, and the instructor prepared me for the conference. The idea was to do the presentation in a TED Talk format, meaning that there would be just a few slides with almost no text, and the presentation would take 18–20 minutes. It looks simple, but I would like to invite readers to try it. The presentation had to be perfectly balanced and made without reading from the slides or other aids. We wanted to engage the audience and get them focused on the presenter (me). I practiced a lot, recording myself many times and reviewing my performance. I wanted it to be as close to perfect as I could manage.

The conference was in Amsterdam, and I was the last speaker. I must say, I was brilliant at the podium (Figure 14.4). My years of experience at AMC, standing in front of the audience and being relaxed helped me a lot—and so did all of the coaching and practice. I was pleased with my presentation, and the audience rewarded me with a standing ovation. After the conference, I received two invitations to make the same presentation: one in Kassel, Germany, to present for the COG (Component Obsolescence Group) and one in Bristol, UK, to present for the IIOM (International Institute of Obsolescence Management). Shortly afterward, I was asked to make the presentation again at Hilton Rotterdam for the automotive industry conference and then for the University of Amsterdam.

FIGURE 14.4 Enjoying speaking at the FOM (Future Obsolescence Management) conference in Amsterdam.

Courtesy of Marijan Jozic.

A year later, the ARINC standard 662 was issued, and I was invited to travel to Dallas to speak at the FOM Conference US. Before Dallas, the organizer provided a PowerPoint designer to polish my presentation and sent me again to my well-known instructor Nick Graham to boost my presentation skills. It was a terrific show in Dallas. Through my work at ARINC Industry Activities and attending many obsolescence events, I became an expert in obsolescence management. Isn't that a great spin-off from my activities? I was really enjoying my aviation career.

15

Avionics Shop

At the beginning of the new millennium, I was working on projects in the KLM maintenance department. Then we got a new VP who had no technical skills at all. What's more, she had a master's degree in political science and was set to run a department that does aircraft C and D checks.

It seemed that the whole aviation ecosystem was changing. Airlines used to have engineers in key positions, and they were both good managers and good engineers. At least they could understand what the engineers and technicians they managed were saying. The new environment was saturated with people who had absolutely no technical experience. This started to become a problem because they could not make the necessary technical decisions. Most were afraid of making the *wrong* decision. To allay this fear, they requested a lot of investigation and support to make a data-driven decision. Many good engineers didn't like the new style of working. In their eyes, a good leader would make the best decision with a limited amount of data (information). The new managers, most of whom were hired fresh out of universities, requested a lot of data, Excel sheets, measurements, and reports before they could make any decision. With that much data, even my great-grandmother could make a good decision. In the end, it came to two options: If you do this, it will cost you that, and if you do it differently, it will cost you less. They would invariably pick the lower-cost option. My great-grandmother can also recognize which number is lower. This was quite frustrating, because the least expensive option is not always the best choice. The situation created a need for decision support managers, advisors, communication managers, process analysts, and, finally, six sigma lean, green, black belts. It started to get complicated. There were too many people in charge and not enough people doing the work. There was also no focus on the core business, which is maintenance or, in my case, modification projects.

Obsolescence

I knew that my days in the maintenance department were numbered. I was ready to go. I used to have such feelings about my workplace every seven years or so. And I already had my seven years in aircraft maintenance. I started doing some projects outside of my department borders and outside my aircraft maintenance VP's area of operation. One such project was a series of meetings with Boeing and Air France regarding obsolescence management. That was an area in which I could use my knowledge about obsolescence. The meetings in Paris were initiated by Air France, who complained to Boeing that over the lifetime of the aircraft, many LRUs become obsolete, and the airlines often don't even know it. When they do learn about obsolescence, it is already late and the repair or replacement is very costly. I went to the meeting with the head of the engineering department. She was known as the manager who changed her hair color every month. To me this was rather funny, because sometimes I didn't recognize her.

In the meetings at Air France, I had to consider the culture. In French culture, the highest manager speaks at the table and his/her staff only speak when asked. Dutch culture is different: Everybody talks, no matter their rank. I decided to respect the French culture, as we were in Paris, and leave it to my manager to take the lead. But at a certain point, I had to jump in and discuss the subject with Boeing and talk to the French team as equals. Finally, we forged the plan. I noticed that my companion (the engineering manager) couldn't even follow the meeting. At one point in the meeting, she asked, "Can we define the process?" There was silence, and I had to save her by explaining that we had just done so. The next day I had to go to her office and explain again that

Boeing would establish the list of obsolete LRUs and fix for obsolescence. They would maintain the list and publish it on MyBoeing Fleet (their website). Airlines can access the obsolescence page of MyBoeing Fleet and see the status of their aircraft type. It was a simple, beautiful plan. Through AMC I pushed Airbus to do the same and establish the obsolescence chapter on their AirbusWorld. I was proud of this initiative.

KLM moved me to become the 747 D check account manager. I was now responsible for D check completion. I had to make sure that D checks were prepared, then follow the six weeks check, report daily to the customer to check the status, and, finally, make sure that the bill was sent to the customer. I didn't like this position because I had to deal with impossible managers. They didn't manage their part of the process well, and every step was always late. Then the company decided not to complete modifications during D checks. I could see weeks in advance that the D check would be delayed, and I was likely the only one to notice.

Avionics Shop

Our HR department used to have a monthly meeting to discuss open positions and see if there were people within the company who could fill them. I sent them an e-mail expressing that I would like to move to another job. Within a week, I got an invitation from the Shop Manager of the Avionics Department to meet and discuss whether I might like to work on that team. I knew the manager from my engineering days; he was messy but technically skilled. After an hour-long meeting, I said, "Ok I am in! But first I would like to go on holiday."

It was summer, and I was on a four-week holiday. Much too long! It took me one week to get used to doing nothing; then, I could relax for a week; by the third week, I was bored. The whole family was happy to go home. I came back and reported to my new job. They forgot that they hired me and didn't even know that I was going to start. There was no computer for me, no chair, no table, nothing. It was August 22nd, and some people were on holiday, so I used their computers and workspaces. I ordered a laptop and got to work in the avionics shop. I had worked in that shop 25 years earlier as technician. In all that time, nothing had changed. What a terrible situation! There was no fancy new tech, the staff was not motivated, but there was a long list of FAA and EASA nonconformities.

Around that time there was an open house for friends and family to visit the workshops and hangars. It was a big celebration as the company was celebrating its 90 years in service. We organized the route through the shop on the IKEA principle. We start at one side of the shop and then went with the flow until the end. Our techs were there to answer questions and talk to the public. Many people who walked through had the impression that they were in a museum. It was supposed to be a high-tech avionics shop and it looked like a museum! This was bad news. I was managing the shop engineering department, and I had to do something.

This was my plan:

- Staff should start to talk to each other.
- Define the work package for each engineer.
- Solve nonconformities.
- Start the project equivalency of tooling and test equipment.
- Launch business cases for new capabilities.

After we start focusing on a common target, we will start really building a team. Once more, I had to start from scratch and make a solid working team.

My biggest worry was that my engineers were in the office with their backs to each other. They were facing the wall and not talking to one another. I moved them to another office and made sure they could see one another face to face to facilitate communication. Still, they would only talk if they did similar work. We had eight shops, and I decided to put the development and production engineers together at the same table block. Because of this, they were forced to talk to one another. They were supposed to make sure that their shop could operate smoothly, and this started to work well. After a few weeks, they became used to working together, and it was very satisfying to see them cooperate.

Nonconformities

My job in avionics was to focus on nonconformities. We had an Excel sheet full of problems, and it was difficult to reduce it to just a few nonconformities. There was one big problem, which was the equivalency of test equipment. It was a costly problem, because we had about 2,000 capabilities, and each of the Component Maintenance Manuals (CMMs) that go with the capabilities had to be reviewed. Management didn't want to spend money for all that engineering work, but Quality Assurance (QA) and the authorities (FAA, EASA) were breathing down our necks. It must be done: No exceptions. We stayed on the project, which lasted for three years. In the end, it was the most enjoyable, relaxing, and inspiring project I ever worked on.

I'll explain how we did it, but first I should describe WHY equivalency. I'll start with the crash of MD-80 of Alaska Airlines Flight 261. The airplane crashed into the Pacific on January 31, 2000, roughly 2.7 miles north of Anacapa Island, California. The National Transportation Safety Board found that inadequate maintenance led to excessive wear and eventual failure of a critical flight control system during flight. The bottom line was that all shops worldwide needed to check the tool used for the maintenance of LRUs. If an alternative tool was used, it should be as good as or better than the tool prescribed. We were an avionics shop, and we had a duty to follow this directive.

My engineers didn't know that there was an excellent ARINC specification for this problem. ARINC SPEC 668 was the answer to Proposed FAA Order 8300.10. The chapter titled Special Equipment and Test Apparatus defines the expectations for airlines and repair stations evaluating equivalencies. This was our starting point. Because we were getting a late start, I was required to send monthly progress reports to the FAA office in Frankfurt.

WHY Setup

I had only one engineer with the skill set for the job, so I had to create a team. One person could not do it all alone: This project was too big. I used to commute to the office by bicycle. Work was only 10 miles from my home. One day on the way home, I saw a familiar face. It was my boss from my earlier years when I started in the engineering department. He actually hired me in 1986. He had retired but did some volunteer work for Dutch Dakota Association. I stopped and asked him if he would like to work on a project. I said that I would pay him and that it would be more interesting than volunteering for that old dusty Dakota airplane. The next day, he came to my office, and I explained the project to him. Just like that, he was hired.

I also knew an engineer who couldn't get an engineering job and was working as a contractor in the incoming goods department, packing and unpacking boxes. He was known as WJ. When I heard that he was an engineer, I grabbed him from incoming goods and got him a job as an equivalency engineer. I now had three engineers. Then I called the Aviation Department at the University of Amsterdam and gave three students internships. It was usually a three-month internship, and it ended at the beginning of summer break. So, I hired those three students for the summer, as well.

We had 2,500 CMMs to do, and the project was running smoothly. We focused on fast movers. Those are LRUs that come to the shop frequently, nearly every day. They were the highest priority. The next group was the slow movers and the LRUs that came to the shop. The equivalency was not done yet, and my team had this as our new high priority. The unit (LRU) would not leave the shop if the tool was not checked and the equivalency report was made.

There was a lot of paperwork because our IT department didn't allow us to use hand scanners. We had to input p/n per p/n manually. This was a boring job for my engineers. Then I had a terrific idea: There was an office at the flight ops where they managed grounded flight attendants. Some flight attendants were not allowed to fly for various reasons: They might be recovering from a broken leg, have excessive jet lag, or be in the final stages of pregnancy, for example. They were willing and able to work, but they were not allowed to fly. I called the office, and they sent me flight attendants for my project. These staff members were wonderful. They were trained to deal with people—that was the easy part for them. Their job was to go to one of our eight shops, borrow three or four CMMs, and then insert the p/ns of the tool prescribed in the CMM in the Excel sheet. And then do the same all over again…and again. They learned how to do it in 20 minutes. After that it all was fun for them. They enjoyed their time in the office and were happy to see the same faces every day. And they did useful work. The whole atmosphere in the department changed, and it was so nice that people started to love their work. I had managed to make a team again.

At one point, we had seven flight attendants and four engineers in the meeting room doing the equivalency project. It was interesting to work with the flight attendants, whose experiences were so different from those of an engineer. They were also very social and would chat while typing those p/ns in the computer. The disadvantage was that technicians from the shops were coming to our meeting room with all kinds of excuses just to socialize with the flight attendants. The operation was successful, and the project was finally completed, thanks to their help.

During this time, a reorganization took place, and a new director arrived. I knew him from my mini cockpit upgrade project for Atlas Air. He was selling the modifications I had developed. I was OK with the new director. He understood that we needed more new capabilities and that those capabilities would bring more work to the shop. We had just developed 30 small capabilities. I managed to buy the weather radar test set capable of testing all Collins WX radars. The others were museum artifacts. The new director knew how to approach this financially, and it was decided to buy the department two ATEC6 test computers. We already had an ATEC4000 for legacy 747 classic

LRUs, but all our 747 classics were gone. Then we had two ATEC5000 computers for 747-400 LRUs and then IRIS and ITS-700 and STS1000. Again, this was all museum stuff. I called the aerospace museum to donate the ATEC4000, but they didn't even want to accept it. They weren't even interested in coming to our shop to see the test set. Another important note about my engineers: They didn't know how to make a business case, and neither did I. We all had to learn it.

Capability Development

In my career, I have visited quite a variety of aviation components shops. Examples include shops that are part of a large MRO, independent MRO shops, small shops with just hub support activities, and OEM shops. Often, when I talk to the managers, they ask, "How are capabilities developed for new components?" Obviously, they do not know the answer, and they do not trust their engineers, because their engineers are telling them that it is a complex process and that it takes a long time to develop the capability. This is always bad news: "It takes a long time!" In my shop it took even longer, because the company did not invest in this process, and therefore, engineers didn't learn how to do the business case or how to apply for capital investment. Managers asked around in an attempt to get the answer they wanted to hear. Unfortunately for them, the correct answer was, "It takes a long time!" My executives visited other components shops to "talk business." Many times, those shops would brag about their performance. On one occasion my manager came to me with this story: "We are slow and inefficient in developing capabilities. I visited a shop in Miami last week and they can develop a capability within three days. Why does it take 8 to 12 months in our shop?" This story was absolute nonsense. You could develop a capability in three days if you were only adding a Phillips head screwdriver to a toolbox or just one more dash number and nothing else. But that is not what we mean by capability development.

Capability development is a very complex process. There is no global standard, manual, or guideline for capability development. The term *4M* (Method, Material, Machine, and Man) is known everywhere, but there is no description or flowchart of the process. In addition to being complex, the process is dependent on individuals or departments to develop capabilities. Every company will develop capabilities differently, and every department within the same company will use its own procedures customized to its needs.

At this time, many aviation companies were closing their capability development departments to save money. Once you lose the knowledge and technology, you are finished. My shop was on the edge of such a situation. There was a lot of pressure to develop the 737 capabilities as the company began operating the 737. Ideally, the shop would need to cover 80% of the flow of components. This is known as the Pareto principle. If you cover 80%, you are covering most of the pool. For the additional 20%, a lot of effort is required for minimal gain. The pressure on us was that our inventory department wanted to negotiate contracts and outsource everything. They always hammered into my head that they didn't have time to wait for my department to establish the capability. Without a contract, they were paying more per event. My shop missed this opportunity in the first five years of operation. The components were under warranty, which is a good time to implement the capabilities. But we were sleeping.

Relaunching a department becomes almost impossible because the people with the specific skills and know-how are often no longer with the company. If there is no written description of the necessary processes, it's like reinventing the wheel. This is, of course, the cost aspect that might prevent the reopening of the capability development department.

Maintenance of components for each new aircraft can be completed in-house, or it can be outsourced. The idea is to analyze the flow of components within the operator's logistics department and set the target to develop capabilities to cover, as

I said, 80% of the flow. Above, I mentioned the 4M method; let's review 4M.

1. Method

The documentation is on how to maintain, repair, and test the LRU. In most cases, that is the CMM, but sometimes you might need other documents such as drawings, Interface Control Document, and test specifications. You should make sure that you get all those documents from the manufacturers. For that purpose, ARINC Report 674: Standard for Cost Effective Acquisition (SCEA) was developed when I was the chairman of that working group. It is imperative that the contract be correct and inclusive so that the shop can obtain all the documents for capability development; otherwise, you cannot go further. Without documentation, there is no capability.

2. Material

We often refer to the material used in the repair process. This is quite a complicated M. You have to decide whether to purchase piece parts or subassemblies (Printed Circuit Boards [PCBs] or modules). For repair Level 3, piece parts are required, and for repair Level 2, Shop-Replaceable Units (SRUs) or PCBs might be needed. This is an important point, because Level 3 repairs are the repairs you want to do in the shop. Those are piece part level repairs for which most parts cost just a few US dollars. It is cost effective, but you still need skilled technicians who can troubleshoot. Otherwise, you move to repair Level 2 and that is on PCBs. This method is faster, but you must send the PCBs for repair, which is costly. The supplier will charge you a lot, and you will need an additional PCB for the PCB pool.

These materials are called *initial provisioning*. The reason to start by ordering the initial provisioning is the lead time. The delivery of those piece parts has the longest lead time. Back in the

1980s, when we did not have so many computers, the lead time for materials was much shorter, because the OEM had everything in stock and directly available. This brings me to another point: having all this stock in stores is calculated as dead capital, so OEMs have changed their strategy. Now the process is delivery on demand, and nothing is kept in stock. When you order a component or piece part from the OEM, they receive your order and then reorder it through their sub-tier supplier. Lead time can extend to 280 days or more, so the initial provisioning should be ordered as a first step after the manuals.

Another method vendors use to sell more parts is requiring a minimum quantity. For example, you might need two transistors per year. If you order two, you might get a message that the minimum quantity is 12. You are forced to buy more than you need to complete your job. Now the dead capital is in your warehouse.

The next step is to calculate how many of each component or piece part your shop will need. This is based on flow. There is one important aspect in this context: Ask for advice from the OEM. Most suppliers are helpful. One I particularly like provided me with a list of piece parts and SRUs they used for repair in one year, which was 250 serial numbers. That was the best list ever! From that list you can easily calculate the amount of material you would likely need for the repair of your flow simply by extrapolating. Another way to tackle this problem is to buy one serviceable LRU (secondhand perhaps) and tear it down it. That way you will have all parts available.

Generally, a shop starts with Level 1 and Level 2 maintenance and later extends to Level 3 maintenance. This is an excellent approach. In such a case you might consider executing contracts for a small pool of SRUs. That will make your TAT (Turn Around Time) more controllable, because you will have the SRUs available. The next step would be to introduce the Level 3 maintenance for the power supply, and once you master that, you can extend it to the circuit board, which is the least reliable.

Materials and logistics can be a nightmare, but you can also consider it a game. If there is time, you can develop models for the calculation of piece parts and make it very sophisticated. Unfortunately, in real life this approach might not be feasible.

3. Machine

The test stand (Machine) is the one big decision that must be determined. The decision is between *make* and *buy*. If your company has skilled engineers in software and hardware, your company might consider building the test stand. I have met several great engineers who could accomplish this task. Such engineers are around, but you may or may not have one on your team.

You are in engineering paradise if your staff has the skills to build the test set. I am talking about a test set for serious LRUs, including the development of a test unit adapter, switching unit, measurement of signals, and so on. It can be a very complicated task, but if management has trust in their engineers, and if they provide full support, amazing things can happen.

Sometimes, it is better to buy the test equipment. It can be Automatic Test Equipment (ATE) or just a stand-alone test set for one p/n. Usually, this option is more expensive, but you get proven technology, and it works from day one. This process includes time-consuming steps such as contract negotiation, business case, and capital investment. It can be a complicated and lengthy process.

My first project was the purchase of a weather radar (WX Radar) test set. It was a useful purchase, but the test set didn't have the so-called CE marking, despite being built in Arizona. "CE" is the abbreviation of "conformité européenne" (French for "European conformity"). The European Union demands that every piece of hardware (toys, test equipment, furniture, everything) have the CE safety marking. There are also other requirements that go along with that marking. My WX Radar had to comply with the low-voltage requirement and EMI requirement. This can be compared with American UL markings (requirements) and American FCC requirements (UL = Underwriters Laboratories; FCC = Federal Communications Commission).

There was little choice but to purchase the test set and do all safety testing and modifications. This immediately created a problem with my management and financial guys. The test set was delivered, but it was not used for two months. In those two months, we sent the test set to an external company for a quick scan. They provided advice, and then my engineers developed the modifications and documented them. Finally, we could order material and modify the set. In those two months I heard complaints that we invested 300K USD for a test set, and nobody was using it. I had to keep the financial team happy and informed, but they tended to share too much information and had contact with the COO. If bad news traveled from them to the COO, he would not likely be cooperative with my team's requests in the future.

The machine for testing LRU can be very expensive and complicated or quite simple and straightforward. If you are developing a test set for an aircraft light, you might need a power supply and a cable. But if you are developing a test setup for an Engine Control Unit (ECU), it is a totally different ball game. Even if you purchase the ATE, you will spend significant time in the correlation and repair source substantiation process before you can do the first ECU. This is a very complicated M.

4. Man

This M might seem simple, but it can be difficult and frustrating. You need to train your employees well. In the 1980s, it was easy. The OEM had a training department that sent the training schedule to the airlines. We knew in advance when and where training was taking place. You could sign up for a training and, in addition to learning necessary skills, you would meet colleagues from different airlines.

Now, suppliers will only provide training upon request. Suppliers will inform you that you are the only one asking for training and, therefore, they cannot provide it. After some negotiation, the supplier will usually provide the training, but for Level 2 maintenance only. This is a continuous game between shop and supplier. It is important to look at the contract and point out the clause in the Product Support Agreement (PSA) to your supplier. This clause says that you have the right to be provided training for your personnel.

There are various types of training, but each has the same precondition, which is workmanship. It is assumed that technicians have basic skills. The training for new capabilities can be defined as follows:

- Similar equipment and no additional training are required. In that case self-study of the CMM will provide sufficient awareness of the differences between existing and new equipment. This is typically the case when only the dash number of an LRU is changed.

- On the Job Training (OJT) is the type of training done by an approved instructor who is knowledgeable and experienced with the p/n. Typically, the instructor is trained by the OEM or another authorized party. Train the trainer principle: this procedure can be part of standard shop practices, and the shop should develop and use an OJT form to guarantee the traceability to authorities. Those forms are filed by the Quality Assurance department.

- The technician is sent to the OEM for training. It is important to train technicians for Level 3 maintenance. See ARINC Reports 663 and 674 for details about levels of training. The MRO is responsible for subcontracting the training for its technicians. It is extremely important to remember that the training levels are in ARINC Report 674. The training levels are defined as per ATA Spec 104 and are in Attachment 5 of ARINC Report 674. There are four training levels, and they do not correspond to the three levels of maintenance. Therefore, you should read ARINC Report 674 twice before you contract the training.

The above description on capability development was the result of three years of struggle with many departments and managers. Eventually, my team learned the necessary skills, and, more importantly, they learned how to do the business case. My team was skilled and operational. In four years' time, we reached the magic number of 80% of the flow. We had an operational shop, skilled staff, a lot of work, and not one nonconformity. Three years after I started as a shop engineering manager the FAA inspector visited us again and expressed his satisfaction in the briefing with the higher managers. FAA was in the shop for three full days doing an audit, and had not one negative remark. My dream team was superb! (See Figure 15.1).

FIGURE 15.1   Me and my dream team in the Avionics and Accessories shop.

Courtesy of Marijan Jozic.

I felt that I had to give something back to my team, so, every year, I made it possible for them to attend a course or join a conference locally. I had 24 engineers, and it was not that big a deal, considering that I could issue a purchase requisition order up to 15K USD. Once, I contracted a "stress management" training for the team. The year after that everyone attended MRO Europe in Amsterdam. The following year they attended a course at National Instruments (NI) on NI products. And the year after that we all went to a SharePoint training. The team appreciated these opportunities, and we all enjoyed the experiences.

16

DORA: Development of Repair Avionics

Shop engineering is very different than systems engineering. System engineers don't have many difficult tasks. They don't have to fight for data and all the knowledge they need; they can obtain it from the aircraft manufacturer. The average engineer can survive in the engineering world without major difficulties. Very good engineers become project managers or department managers. Nowadays, companies often have non-technical people manage the shop. These managers tend to have a very good education but no technical knowledge. This makes communication between them and the engineers difficult.

The shop environment is a special piece of cake, and shop engineers—avionics or hydromechanical—are a different breed. They need a lot of knowledge about measurement and regulation techniques and a lot of general knowledge about electrical engineering and physics. Most of the time it is better to have an electrotechnical engineer in the department than an avionics engineer. Further, there are two possibilities. There are shop engineers who can do the business case and purchase the test equipment but are unable to design the circuit chart or write the software code. A manager can consider himself blessed if he has a few engineers who can design the PCBs and/or the whole test system. The most skilled are the engineers who can do all that *and* write the software code. Such engineers are worth more than their weight in gold. I was blessed to have a few engineers who could do the business case, design the test systems and circuit boards, and write the software code.

After many reorganizations and budget cuts, I found myself in a department where engineers were supposed to sit behind their desks and produce paperwork. There was no place where they could experiment, try out their designs, or modify test equipment. It was not possible and practical to do in the shop on the production line, nor was it possible to do in the office. We had a small room with some tools, but the demand for test equipment was high, and I had to do something. My first idea was to check what the bleeders were: the problems my engineers were facing every day that were making their life miserable. My engineers were not delivering capabilities, and I was sure that there were more reasons or a combination of reasons for this poor performance. I could talk to them in a "one to one" meeting or just try something different. My idea of "different" was to invite them for one full day to a meeting in a separate building where we stayed until we forged a plan. It was a challenging idea. It was me against 24 engineers. That could lead me to 25 different opinions and no conclusions. The team members were very cooperative, and these were the most important findings:

1. When they start to design the hardware, they need a lot of dedicated time, and they don't want to be disturbed with day-to-day questions.
2. When they start to write software, they prefer to be left alone for days without being disturbed with day-to-day issues.
3. They want to be informed about the current situation in the department.
4. They wanted an easy procedure to be able to purchase small tools and equipment.
5. They need knowledge about CE marking and programming of NI PCBs and devices.
6. They need a dedicated area to be left alone and without interruptions.

Believe it or not, all of those items are reasonable—not only reasonable but the lifeline of the engineer. After that meeting, I had our To Do list. It was a joint effort, and we all knew that if we managed to solve those issues, our department would perform better, and we would all be happier.

Items 1 and 2 were easy to fix and involved the work distribution between engineers. For each subject I appointed a lead engineer and a deputy. I was backup for them for all systems and subjects.

If lead engineers are writing the software, all questions are directed to the deputy, and all overflow to me. They were in a rather luxurious situation when their boss is able to handle all the subjects if required. That is how we isolated the engineer and enabled them to do the programming or design of the test setup. That job didn't last more than two, sometimes three, weeks, but even if it was four or five weeks, we could manage the situation. We knew that we were helping each other, and we were doing it for the team's benefit.

The idea of informing staff about department and company news had crossed my mind before. At systems engineering, we used to have a binder where the department manager inserted newsletters, general memos, and magazines. He used to send the binder around, and we circulated it among engineers. I tried that method without success. I had more engineers, and it took three to four weeks to circulate the binder, so the news wasn't new by the time it made it to all engineers.

My next quite simple idea was to have a department meeting every Friday after lunch. I called that meeting Sisyphus (Figure 16.1). Why Sisyphus?

FIGURE 16.1 Every engineer is Sisyphus.

vchal/Shutterstock.com.

Sisyphus

In Greek mythology, Sisyphus, the king of Ephyra, was sentenced to an eternity of frustration and wasted efforts because he believed his cleverness surpassed even the greatest of the gods. Every day, Sisyphus would roll a huge, enchanted boulder up a steep hill. Before he could reach the top of the hill, the massive stone would roll back down, forcing him to begin his efforts again. As he pushed the rock up the hill, it occurred to him that life was a journey, not a destination. This story is more than 3,000 years old. There is symbolism in stories, and symbols have different meanings for different people. This is my understanding of the secret code of the Sisyphus story.

My engineers were pushing the rock up that hill the whole week. On Friday the rock might be at the top of the hill, but on Monday they would find themselves at the bottom of the hill again, and they would begin pushing that boulder back up the hill. It might be a different rock or some other weighty matter, but the uphill battle was the same. I am convinced that Sisyphus was an engineer.

I will now deviate from the storyline to explain the Sisyphus problem. Once upon a time, there was an avionics engineer working in an avionics shop. The company had purchased a fleet of 787 aircraft. The avionics engineer received a list of p/ns from the airframer. The list was called RSPL (Recommended Spare Parts List). He isolated a few p/ns and decided to start capability development.

Manuals

The engineer's superiors asked for the plan, and based on his experience with legacy aircraft types, he said, "We should implement the 4M. Let's be careful and add some time to the process of capabilities development. OK. It will take six months per capability. Since I cannot work on my project 100% of the time, I can do four or five projects simultaneously. In six months, the capabilities will be up and running." As he had done before, the engineer sent an email to his friend at an OEM asking for a CMM (Component Maintenance Manual). The response was, "I have no authority to send you the CMM because it is our intellectual property. Our tech pub department will help you." The engineer contacted them and received a short email with an attachment called an End-User License Agreement (EULA). If he signed this and sent it back, he would get the CMM. OK, this was a setback. He had to go back to his legal department because the EULA was full of legal language, which was written in a very small font. Legal checked the document and said, "No, we cannot sign this. We need to change the wording in some places." They made the required changes, and the engineer returned the EULA to the OEM. Now their legal department said, "No, we do not agree. We want it worded differently." After a few more rounds of legal back and forth, the EULA was signed. Then, after some more time, the CMM was delivered. Our engineer immediately noticed that the CMM was very thin. The CMM was not useful, but it referenced a Technical Support and Data Package (TSDP). This was another setback, because the engineer had to go back to the OEM, get another EULA, send it to legal, and so on. After four months he received the required data. One *M* is done, just three more to go.

Of course, our engineer is pretty savvy. He did not sit around for four months, waiting. Instead, he was working on the other *M*'s.

Material

Material refers to initial provisioning, that is, material required in the store when the shop is doing maintenance and repair. Our engineer is supposed to analyze the IPC of the unit and select the parts to be purchased for the first shop visits. He can buy all the parts, but there is a chance that some of the parts will never be used. If he reads the contract, he can find the clause that gives him the right to request the list from the OEMs with recommended piece parts. If he asks the OEM for this list, they will typically say, "No, we do not have it." Another setback. It takes some time before the OEM begins to cooperate. Then he receives the list and a notification that their lead time is 36 months to deliver the parts. One more setback. After some time, our engineer decides to purchase the unit from a secondhand market and perform a teardown. He will obtain all the parts, which he expects much more quickly than 36 months. In theory, he can repair only one LRU, because he has just one of each necessary part. He will still have to order some parts and wait 36 months. This is yet another setback.

Man

This used to be easy. You would ask the OEM representative when the training was being held and you would send your staff to the course. Now, it's different. There are no scheduled courses; everything occurs upon request. And even worse, the courses are very short: two days, three days if you are lucky, but not more. We are dealing with extremely complicated avionics LRUs, and the course is just two days. How much can you learn in two days? A short introduction? Theory of operation? A quick peek into the books and schematics? Maybe a visit to the repair facility, and if the avionics engineer is lucky, he will see the LRU. This is a potential setback to be tackled later. After the unit is in his shop, the engineer will have to learn the hard way.

FIGURE 16.2   Inspecting soldering at a printed circuit board.

Courtesy of Marijan Jozic.

Machine

The machine is the test set. If we are talking about a 787 test set, then we are facing one setback after another. It takes six months or more to get a quote for the test set. When the quote is provided, it is incomplete. Some items are not quoted, some items are missing, and some items are not required but are quoted anyway. So there are a few setbacks before the quote is complete. The next surprise is the extremely high price. The price is called a "quote to kill." If our engineer survives his attack of anger and outrage, he can calm down and look in the books to figure out whether he can design the test set in his own facility. Can you imagine? Talk about a setback! But the avionics engineer is passionate and he knows he can do it.

Eventually, he will see the light at the end of the tunnel. Of course, there will be a series of setbacks in the development process, but a year later the test set is operational. The material is in his store, and the technicians are trained. The latest revision of the CMM is on the shelf, and the engineer thinks he is done. The only item not onsite is the LRU. Our engineer realizes that the LRU is outsourced under a long contract (typically 10 years) and that there is no way to get it in his shop. He was dealing with purchasing, trying to get the parts, training test sets, and other items ready, while another staff member in the inventory department was dealing with other purchasing people who were writing contracts for outsourcing. This is a *major* setback. But the engineer digs into the data and contracts and discovers an opening in the contract: If the shop established the capability, it might be possible to deviate from the contract upon mutual agreement. The OEM is, of course, unwilling to implement this section of the contract. They admit openly that they are going to lose revenue if they allow him to do the repair in his shop. After a few more setbacks, a change is made to the contract, which is signed by the relevant people, and the engineer is ready to go. He may realize at some point that the LRUs have still not arrived at his shop. The computer system is not updated with the latest data, and the LRU is still outsourced. After this setback, the LRU finally arrives at the shop for repair. Instead of six months, it is now two years of lead time for capability. The engineer's name should be Sisyphus.

Let's count all the setbacks: there are approximately 13, not counting a few *series* of setbacks. The engineer is rolling the stone uphill. When he thinks that he has made it to the top, the stone rolls back down the hill (sometimes all the way down). Sisyphus was punished by being forced to perform hopeless, endless labor. The engineer is also punished with hopeless labor (though maybe not endless), and he knows that many of his actions were unnecessary 10 years ago. The avionics engineer is conscripted until his retirement, which, while not endless, is bad enough!

My weekly Sisyphus meetings were important. We could exorcise each engineer's issues, share experiences, and help one another push the rock up the hill for another week. It was a very well-attended meeting, and my engineers were always on time. If the meeting took longer than an hour, nobody complained. It was nice to talk to others fluent in the engineering language. Back when I had flight attendants in the department, they sometimes also attended the meeting as part of my team. But after the first meeting, they admitted that they didn't understand what was going on. The engineering discussions were too complicated, and they got the impression that it was a different language full of acronyms (they were correct). The atmosphere in my department was improving and that was a good sign.

Purchasing

The next issue we had to resolve for the team was that they wanted an easy procedure for purchasing small tools and equipment. This was complicated because I had to negotiate with management. I wanted to give my team members purchasing power. This meant that they would be able to issue the purchase requisition order. This resulted in some management drama that ended with the words: "But now you also have the purchasing *obligations!*" Management felt that, if the shop engineering team was to have purchasing power, they would also have to negotiate with the supplier's sales people. This was not a big deal. Our purchasing budget was never more than 10K USD, which is a small amount. If the limit was greater, a different internal process would be required. The only thing my engineers wanted was to be able to send the schematic to the supplier who would design and build the PCB. This is sometimes very technical, and the cost can be a few thousand USD. Sometimes they simply needed to order wire or a set of screwdrivers or a label printer for 400 Euros. This is what my engineers asked for; this is what they got.

CE Marking

The next request on the list was for knowledge about CE marking and programming of PCBs and devices. The team obviously lacked knowledge. CE marking is the European law about the safety of equipment and tooling. Everything in the workshop produced or purchased after 1996 was required to have the CE safety marking. We didn't know what the rules and regulations were or who could mark the equipment. This was the perfect project for an internship. Because I had a good relationship with the University of Amsterdam, it was very easy for me to get a responsible student to investigate the CE marking you can find on every device built and used in the EU. The odd thing is that we could purchase the test equipment without the CE marking, but in that case, the EU law would hold us responsible for the CE marking, not the manufacturer.

FIGURE 16.3 Safety markings.

I visited the Dutch Aerospace Laboratory with the student intern. I was a member of their advisory board, and they were more than happy to help me. We learned that if we purchased something with UL and FCC markings, that would be equivalent to a CE marking, but we had to train our staff and make a technical file for each case. If we build something in our own facility and we respect the EU regulation, we must also mark our equipment with the CE marking. The student made a policy document, and I hired an instructor who trained my team about CE marking. We established workplace instructions for the process. It took almost half a year, but, finally, we were in business.

If you are brave, luck will be on your side. Or so an old saying goes. With respect to NI (National Instruments), it was easy. They have free courses for their customers. The course was held no more than 50 km from our shop, and my engineers could go in groups of three. If I had made it official, many managers would ask questions and a year would pass before the first engineer could attend a course. All my engineers were scheduled for training, even if most of them could not do the programming. The idea was to send them to the course to give them a feeling for what can be done with NI equipment. That is the first step toward finding a possible solution to their problem.

Dedicated Area

The last issue, in which the team requested a dedicated area, was a tough nut to crack. We didn't have many capabilities—actually, the capabilities we had were for aircraft we were not flying anymore. There was a lot of legacy equipment that was no longer used, but which nobody wanted to scrap, because it was still in the financial books. My first step was to start scrapping unused and outdated equipment. There was a switched-off ATEC4000, Bendix DC10 autopilot test set, and what felt like millions of small test sets. There was also a clean room of the highest class, but not one LRU was maintained in such a way that the clean room was required. This was a very big area in which only four people were working.

They were convinced that a clean room was required, and they pointed to the CMM, which listed "clean working area."

They had been able to sell that story to previous managers, but not me. It was a business case of one page, but it was enough to shut down the clean room. The area was degraded from a clean room to just a normal avionics shop. The savings on filtration units was 24K USD per year. Because it was the biggest shop area but it was nearly empty, my colleague managers used the opportunity to reorganize the shop. They threw out all unused test equipment and moved three small departments together into that huge area. One of the smaller areas remained empty, and I planned to confiscate that area to build the lab for my engineers. The idea was to have four stations where my development engineers could work. They would be able to build the test set based on NI equipment and do programming and experimenting. After a test set was finished, they would deliver it to the shop, train the operator on how to use it, and start a new project. The area would only be accessible to my engineers. They could leave all their papers and tools on the desk and come back the next day and carry on with their work.

After a tough negotiation and many emails, meetings, memos, and promises, I got three-fourths of the area I coveted. One-fourth had to be reserved for small materials storage. I had to build a wall to separate that area. This was a setback. If I were a demanding jerk, I would probably have been able to get the whole area, but I am too nice for such underhanded behavior.

ATEC6

During this time, I was managing another problematic project. The company had purchased two ATEC6 series test computers. These were in an area with a weak air conditioner. Because of heat production by the ATECs and the temperature chamber, the room got overheated, and our test results were out of tolerance. We didn't know about this issue until those ATECs were installed.

I had to go back to the managers to ask for extra money for an additional air-conditioning pack. I wanted an additional 20K USD for the heat exchanger to reduce the heat and also heat the water we used in the restroom when washing our hands. The building was built in the early 1970s, and nobody ever thought to install a boiler to produce warm water for the employee restroom. The deal I made was that I would get the room for my engineers if I dropped the heat exchanger requirement. At least somebody would be able to say that they saved the company an investment of 20K and that employees still wash their hands in cold water. Installing an electric boiler was not even considered. The air-conditioning was blowing a lot of hot air through the roof of the building and heating the streets and atmosphere. I know that the ROI for the heat exchanger is long, but I was convinced that it would be a subject of bragging rights for my managers in front of visitors: "Look! We use the heat produced by the ATEC6 to heat the water for our employees to use in washing their hands!" But unfortunately, no dice, and no warm water for employees.

My engineers got their lab, though. They had millions of ideas, and I didn't want to interfere with their ideas by organizing and dividing the workspace. I found that many people in the office were jealous and tried to make things difficult for me whenever they could, just because I was able to find a creative solution to meet my team's needs.

DORA

In one of our Sisyphus meetings, I asked my engineers if they wanted to give a fancy name to the area so that we could put a sign on it. After some discussion, the decision was made to call it Development of Repairs Avionics, using the acronym DORA. I liked the name because it was different and that made an impact. Also, if you give something a name it seems more official. I used to run a modification program with the number Modification Order 34-750. It was a big project, but nobody talked about it. In contrast, my fellow cabin engineers designed an ordinary shelf for the galley in the 747. The purpose of the shelf was so that, during a flight, before the crew retired to the crew rest area, they could pull down the shelf and put cans of soda and bags of pretzels on it. The passenger is then free to take a can of soda and a bag of pretzels if they desire and the crew is not bothered. They called that very simple modification "Project Iris." It was just a wooden plank with a hinge, but the whole company was talking about the fascinating Project Iris. All of top management was aware of Project Iris.

DORA was a great name. Of course, I didn't know about *Dora the Explorer*. *Dora the Explorer* is an American children's animated television series that premiered on Nickelodeon. The series focused on a girl named Dora who was accompanied by a monkey called Boots. They would go on adventures and solve numerous challenges along the way. There is some symbolism here with respect to repair developments, but I was never sure if I represented that monkey. We ordered a proper sign for the entrance to our lab with a picture of Dora and the text DORA—Development of Repairs Avionics (Figure 16.4).

FIGURE 16.4 DORA: Development of Repairs Avionics.

I fulfilled all the wishes of my engineers, and now it was their turn to deliver, which they did. The record was to develop 24 capabilities in one year, which is a lot.

As I mentioned previously, I had a difficult time convincing my management about the DORA lab. Sometimes I asked myself, am I crazy to spend so much time and patience convincing my unsupportive management just to get this lab? It was not a pleasant experience, but management had to make decisions without any idea what capability development or repair development means. Their attitude quickly changed when they noticed that our customers, visitors, and colleagues wanted to see DORA. Almost every week there was someone knocking at my door asking for a tour through the shop, especially to see DORA. The same managers who gave me such a hard time were now bragging about DORA. Suddenly, DORA was their brilliant idea. What should I say? I guess *pathetic* would be the best word for it.

17

AKCD: Aviation Knowledge and Career Day

Established in 2003, there was a whole chain of yearly avionics conferences in Amsterdam called Avionics Expo. I found myself on the steering committee of the conference, though I still don't have any idea why and how I landed there. Before the conference, we had a few meetings with the goal of selecting the presentations. The setup was such that there were booths in the big exhibition area and a conference room for seminars. One year, the organizers offered me a booth for free, and I happily accepted. My idea was to get a few of my engineers in the booth with a test setup that we designed. That was a motivation for my engineers, and at the same time, a way to reach potential customers. This was a mistake on my part, because I got the whole sales department on my back. Sales would usually have a booth at conferences, and now the avionics shop was invading their territory. I had a few flight attendants in my department at the time, and I asked them if they would like to spend a few hours in the avionics booth wearing their uniforms. They happily accepted the invitation. This was an even bigger threat to sales and marketing. Eventually, the COO decided to let me go to the exhibition event no matter what the sales team said.

We had a great time, and the biggest gain was that it was very inspirational for the department and the shop. They had a manager (me) who was able to do more than to keep them in the dark and between the four walls of the building. They were motivated to do more, and I was not even asking for more. The event lasted only two days, but it was educational for all of us. I was in the booth most of the time, and at one point, a person approached me with a story about "Career Day" at the Amsterdam University of Applied Sciences. He introduced himself as Peter and said that he had heard about me and wanted to invite me to speak at the event. I asked, "What is it about? What subject do you want me to handle?" He didn't care as long as I agreed. So I agreed, even though I didn't know what I should present.

A few months later, I was at the podium in the amphitheater of Amsterdam University of Applied Sciences. In front of me was a group of 300 individuals, including students, teachers, instructors, engineers, and some officers of the Royal Dutch Air Force. My presentation was about managing the LRU's within an airline. For me this was not an exciting topic, because I had been in that environment for 25 years, but they seemed to like it. At the end of the day, Peter, the same gentleman who approached me at the Avionics Event asked me if I would like to join him on the steering committee of AKCD (Aviation Knowledge and Career Day). There would be two meetings per year, one dinner before the next AKCD and one full-day (Saturday) meeting at the Amsterdam University of Applied Sciences in March each year. Why not? It might be fun—I was already enjoying it. It would be good to meet other people involved in aviation in the Netherlands. I had been in aviation for more than 25 years, but most of my contacts were abroad, despite the Netherlands being my home and a country with a lot of aviation history.

AKCD Steering Committee

After my first AKCD in March 2003, I didn't hear from Peter. I pretty much forgot about the event. He called me in the fall of the same year and invited me to a pizza dinner with the steering committee. We met in the old part of Amsterdam. I had never been there, but it was not difficult to find. I wasn't sure I was at the correct address, but when I asked at the bar about Peter, the bartender told me that Peter had reserved a table where I could wait. Within a few minutes, the steering committee had arrived. There was a gentleman from Air Traffic Control, an officer from the Dutch Air Force, an owner of a flight attendant training company, an engineer from the Dutch Aerospace Laboratory, a journalist, me (an engineer) from KLM, and Peter, of course. Peter was a pilot at Air Holland, a small airline that had operated for just two years. During dinner, we talked and got to know each other. After pizza, Peter started the informal meeting. Everyone threw out ideas about topics for presentations for the next year. In the end, we had an impressive list of 25 presentation ideas. Peter made clusters of those topics, and after that, we started proposing speakers. In 1.5 hours, we had defined the whole day's schedule: clusters, topics, and potential speakers. Our duty for the next weeks was to contact those potential speakers and reserve them for the next AKCD, which is held the second week of March.

Working with this group made me realize what the term *professional* means. It was a pleasure to be part of this committee. Each of us knew exactly what to do and how to approach the speakers. Generally, it was not difficult to get them on board, but first, you had to find their telephone number and email. This was most difficult when it came to the big shots—the company VPs and COOs we wanted reserved as keynote speakers for the morning or afternoon sessions. Unfortunately, these people were very well protected by their administrative assistants. The admins were instructed to prevent any direct contact with them, and we received the same responses every time:

- The COO (or VP) has too many engagements, and, therefore, cannot accept the invitation.

- The COO works so hard the whole week, and his/her weekend time is personal. He/she needs it for family time, and, therefore he/she cannot accept.

- The COO is asked so often to speak at events that he/she could speak every day at some conference. His/her schedule is full, and therefore, he/she cannot accept.

These excuses might all be true, but they are also tiny demonstrations of power. These industry leaders wanted to be begged and made to feel essential. Certainly, they may wish to keep their weekends free for the family. But they also must consider the need to sell their organization and their business to the public. Aviation is competitive, and these leaders will need those young people from the university to run their departments one day. This is the whole point of knowledge and career day: Inform students about real life in aviation; tell them how tough it is; tell them that you care; explain what your company needs. Excuses about hundreds of engagements and family weekends seem like arrogance to me. The COO should always make time to speak for 20 minutes to a group of interested students. Honestly, such opportunities are rare. The date for such an event is known at least three months in advance, and if there is a real conflict, the COO can send a deputy to represent the company.

AKDC Speakers

From my perspective, AKCD was the best way to spend my free time. It was always fulfilling, and I learned new things every time. Why? I met some truly remarkable people, including:

Heather, a 787 test pilot: Boeing was in the process of testing a 787. Heather, the test pilot, came to Amsterdam to brief us about various tests Boeing was doing to certify the 787. She showed some interesting videos about test flights that were not accessible to the public. Once, Heather presented from Seattle. It was 3 o'clock in the morning for her, and it was before we had Zoom or Microsoft teams. She had sent the video clips via email in advance, and she made the presentation over the phone. It was a terrific show.

Christian, an A380 test pilot: He provided a lot of insights and many short videos about the test program for the A380. Christian was one of the test pilots at Airbus, flying different types of planes, and he shared his fascinating experiences.

Fokker chief engineer Rudi, a living legend: Rudi was the chief engineer at Fokker for Fokker F28, F100, F70, and MDF100. The MDF100 airplane was a joint venture between McDonnell Douglas and Fokker. Unfortunately, that airplane was never launched, but the F100 was a very successful airplane that many airlines were flying. Rudi worked on the F100NG concept after his retirement. In his briefings about F100NG, we learned that building the airplane was a compromise. After Fokker's bankruptcy, some companies wanted to relaunch the F100 production. For that purpose, the airplane had to be upgraded. The situation was such that Rolls-Royce didn't want to participate. Therefore, they had to choose a different engine. They chose a Mitsubishi engine that was 500 kg heavier. The two engines are on the back side of the airplane at the tail. To keep the aircraft balanced, they could put bags with sand on the front side of the wing or extend the fuselage on the front side of the wing. The good news was that they could put more seats in the aircraft, but the bad news was that they would be in the same category of aircraft as the 737 and A320, which is very stiff competition. This project required some serious compromise.

Every year one of the guest speakers was a professor from the University of Twente. He kept the audience engaged with a view of his analyses of a situation in aviation every time he spoke. One of his most engaging presentations was about flying an aircraft when a volcano in Iceland erupted. This presentation was riveting!

At AKCD, we had a cluster that was completely organized by military people. They provided presentations about helicopters (Apache and Chinook) and F16 and KDC-10 tankers. On one occasion, their boss, the commander of the Dutch Air Force, visited the conference. After lunch, he came uninvited to the podium and said, "I came here today to see what my [staff] are doing. And because I am here, I wanted to say something…." And he gave a short speech about the Dutch Air Force and about making decisions. It was a remarkable appearance; he was very authoritative. Afterward he left the meeting and left the attendees flabbergasted. A few years later, at the reopening of the Dutch Aerospace Laboratory, I met him again and decided to say hello. We talked for an hour about all kinds of things. I saw him in a different light: He was no longer a powerful top general but a very pleasant conversationalist.

The briefings and presentations at AKCD were wonderful, inspiring, and a unique opportunity for students to learn about the reality of a career in aviation.

Organizing AKCD

Back to the beginning: After that pizza evening, we went to work recruiting speakers. After they were booked, we asked students in the marketing and communication department to develop the webpage and advertise the event. The day before the event was the speaker's dinner. It was in a restaurant in Amsterdam, and after every course, we changed places at the table to mingle as much as possible and talk with all the aviation enthusiasts. During the first course of the meal, you might find yourself near a pilot, spend the second course near a university professor, and settle in for the third course beside a company COO or VP. This was great, because there was usually no time to get to know one another at the event. Steering committee members introduced the speakers and served as their wing mates during presentations. After a full day, there was a closing reception in the lobby, and the event was adjourned.

When I started on the AKCD steering committee, my oldest son was starting his course of study in chemistry at Leiden University. He told me that when he finished his studies, he would not go to work in an aviation environment. He felt it was too uncertain, and there are too many regulators and rules. After he graduated, he got a job at AkzoNobel, a large multinational chemical company. And guess what—he was assigned to the research and development department for aircraft paint. At one AKCD, I introduced him as a speaker (Figure 17.1). He made an informative presentation about developing aircraft paint.

FIGURE 17.1   My son, a chemical engineer, presenting at the University of Amsterdam during AKCD.

Courtesy of Marijan Jozic.

I was concerned that the topic of paint would be an area of little interest at AKCD, but he had a full amphitheater of very enthusiastic audience members. Perhaps the AKCD got him over his refusal to work in aviation. He is now a process engineer at KLM. Aviation, it seems, is no longer such a complicated and unreliable environment.

Spin Off

I spent ten years on the AKCD steering committee and met many interesting engineers from the aerospace industry. After my first AKCD, I also learned something very useful for my day job: I was able to recruit students for internships at KLM. This was a superb fringe benefit. Those AKCD days were very interesting for students, because they could learn about many aspects of the industry from industry professionals. Often the subjects were not part of their university program but were topics they would face when they entered the industry. For me, it was easy to get a good student for an internship. I had my own department, and every application was approved by my management. My idea for the internship was to let the intern do the project for the university in the first two months, and after that, I would treat them as an engineer and let them work for the department for two months. My students were doing department policy papers for CE markings, MSD (Moisture-Sensitive Devices), EMI (ElectroMagnetic Interference), lead-free soldering, and so on. I was their supervisor and could keep them from overloading the team engineers. Once that "official" work was done, the interns helped the engineers on their projects.

I didn't call them by name, but rather "intern"—and they were proud of the designation. On the last day of internship, I released them officially from the internship, gave them a present (which was always an aircraft model), and also gave them their own names back. They had to go back to university for half a year and then were supposed to do a big project for graduation. Some of them came back to my department for their graduation project, which was quite a burden on my time. I had to supervise them and be present at the university when they had their final exams. We later hired a few of my best interns, and they are still at KLM. When I meet them, I still call them "intern," and they are still proud of it.

Once or twice every year, the university asked me to do a two-hour presentation for third-year students. I always enjoyed these. In the beginning, I thought, "Two hours, would I be able to talk two hours to students?" All my presentations were balanced, and I have a number ready for various engagements. I have 20-minute presentations about various subjects, and they are in the TED Talk format: 20 minutes with few slides and not much text on the slides. These presentations are more extemporaneous: No reading from the slides, or even looking at the slides, and it is 15 slides in 20 minutes! After some trial and error, I found that

I could deviate from the slides if I wanted, because the audience never knew what the full content of the presentation was and couldn't tell when I was improvising. Once, after making a presentation for the third time at a different venue, I overheard a conversation in the lobby. One person was telling another that this was the third time he had heard my presentation about obsolescence, and every time it was different. That encouraged me to use anecdotes and stories from everyday life in my presentations, especially when they were longer than 20 minutes.

After 10 years of working on the AKCD steering committee, the university decided to run it on their own. Previously, they had facilitated the event by providing three amphitheaters and parking space. The seven steering committee members did the rest. The university politely asked us to step back from our duties. The AKCD was attracting hundreds of people, and they wanted to turn it into Amsterdam Aviation Week. So we stepped back. A year later nothing happened at the University of Amsterdam. There was no AKCD and no Amsterdam Aviation Week.

This experience shows that when you get a team of motivated people together, you can accomplish wonderful things, and everyone can learn something. AKCD ran very successfully for 10 years. It is easy to stay at home for the weekend and post likes on Facebook or just do nothing at all. But to organize an event that can draw 300–400 people is a very rewarding accomplishment. I was not paid for my efforts; I even had to pay for my own pizzas! But for me it was a fantastic experience and a terrific networking opportunity. I met many interesting people and became good friends with many engineers. I also helped many students become engineers. For some, I was a role model, and they wanted to be a cool engineer like me.

18

The Human Face
of Six Sigma

Today, most people have heard something about six sigma. Six sigma has become a religion that was invented by one of the world's major manufacturers. There is nothing wrong with six sigma. The inventors are smart enough to sell the six sigma concept to believers and to people who want to use six sigma to gain personal success by kicking the butt of decent aviation engineers involved in maintenance and engineering

In this chapter, I'll review some six sigma actions and blunders. This may be bad for my image, but somebody must speak up about this development. Note that the sentiments expressed in this chapter are my opinion only. You can be your own judge, and you have every right to disagree. It is my nature to challenge. Challenge keeps us sharp.

Six Sigma

Six sigma is presented as a set of tools used to improve business processes and quality control. It was developed in the 1980s by a Motorola engineer named Bill Smith. There are a lot of myths about six sigma, but the theory says that the normal production process would deliver 99.5% of good products and 0.5% will be discarded (scrapped) as bad. The outcome calculation is based on three standard deviations. This could be called three sigma, but nobody calls it three sigma (except me).

The "smart" people were not satisfied with three standard deviations, so they established the new norm of six standard deviations (therefore, they call it six sigma), which basically means one failure in 3.5 million products. This is basically the six sigma theory. See Table 18.1 for the six sigma levels.

TABLE 18.1 Six sigma levels.

Sigma level	% of good products	Failures per mil products
1	31.0000%	690.000
2	69.1000%	309.000
3	93.3200%	67.000
4	99.3790%	6.200
5	99.9767%	230
6	99.9997%	3.4

© SAE International.

Everyone in business is doing their best to get all processes under control to prevent failures. The results are amazing at Toyota, General Electric (GE), and some other production companies. For Toyota, it is important to get the production line accurate so that only one of 3.5 million cars will have to be scrapped at the final test. Can you imagine how much money is wasted in auto production? In fact, it's almost nothing compared to the produced quantity of cars. Toyota could afford to put that one bad car in the press and destroy it completely knowing that the 3.5 million cars produced after that would be delivered without any mistakes. Isn't that great?

It is the same at GE. They can produce compressor blades for the jet engine almost without a scrap. Fantastic! That is the glory of six sigma. Of course they are making a religion of it. Wouldn't you? The six sigma priests are not called priests, bishops, cardinals, or popes. They are called lean belt, green belt, black belt, and master black belt (the last one is almost as good as a master Jedi; do you remember Yoda?). Their knowledge of processes is very good, and their knowledge of statistics is excellent. They will tell you exactly how important it is to control your processes and establish a flow through the factory. After that, you must focus on other smaller enabling processes.

You can, of course, roll out this philosophy in the repair shop. The techs in your shop can be compared to surgeons. Engineers are there to solve all the problems to keep the flow going, and the surgeons perform the operations. Flow is the most important part, and everyone must run to keep the flow going. So far so good.

Those master black belts (Jedi) will give you great presentations on the theory of six sigma and their "war" stories. They are so good that they can come to a factory that is performing poorly and turn it around. This is what they call a "fix" or "close" appointment. They will define the critical Xs to simplify the production process, establish flow, lay off some (or a lot of) employees, and book the success. Or they can shut down the plant and save money that way. The audience, people trying to qualify for one of the belts, will, of course, applaud and hope that one day they can also become a black belt.

But here is the catch. All six sigma success stories describe the production environment (Toyota, GE, etc.). No one has provided a story about a low-performing avionics shops that was changed into a great profit center after having been visited by a six sigma master Jedi. Why? Because this story doesn't exist.

The Avionics Shop

Unfortunately, there is a lot of baloney in the six sigma religion. The avionics environment is very complex. We are talking about repairs, and the repair environment is much, much more complicated than a production environment. Nobody wants to admit this; they prefer to think that six sigma is a one-size-fits-all fix that will work in any environment, including the avionics shop.

Let's start with the input. A decent avionics shop will receive 100 LRUs per day. Not 100 parts of one single type or p/n, but at least 75 different p/ns. Shops with more capability will have more variations. Not every technician is able to repair every type of unit, so you need a variety of specialties in techs. Not every LRU can be checked on the same test set, so you also need a lot of different test sets. Not every LRU can be tested with the same CMM test procedure, so you need a variety of CMMs. You need a lot of different parts for those LRUs. Therefore, a good store should be supported by the material planning department to get you all those parts on time. Here we are—people, material, manuals, test sets. You must have all of these for a few hundred or a few thousand scenarios. This doesn't include the modifications that are running through your processes, troubleshooting of home sick units, development of capabilities for new p/ns, and solving nonconformities found in numerous audits. And there is also the equivalency of test equipment and updating of CMMs and software.

It's extremely difficult to keep all these items under control, every day, all the time. Unfortunately, the six sigma master black belts don't like too many variables. Too many variables make it difficult for them to get the process under their control and could lead to failure. To many variables make them nervous. There is a gap between theory and practice. The six sigma black belt is very knowledgeable in statistics but has no clue about avionics equipment. For the black belt, everything is one big mess of avionics stuff. They are not solving technical problems for the avionics shop but rather are annoying technical staff with the six sigma ideology. When they talk about visualizing production, what they mean is that pragmatic technicians should shift 3M Post-it notes along the wall displaying the changing production steps. The walls are then full of those notes (Figure 18.1).

Give me a break. I want to repair the rotables. I don't want to waste my time explaining to highly educated and extremely

FIGURE 18.1   Well-known six sigma scene.

Andrey_Popov/Shutterstock.com.

well-paid statistics people that I need a decent oscilloscope to repair the LRU. It's frustrating when they ask what an oscilloscope or an LRU is. The black belt is focused on measuring, analyzing, and changing the processes and presenting their analysis in beautiful graphs. The problem is that they are speaking a completely different language than the avionics maintenance technician.

In the production department of a Passenger Control Unit (PCU) manufacturer, there are far fewer variables. You can pick up a person from the street and teach him how to assemble the PCU, and in 2.5 hours, you will have a production employee on the PCU assembly line. Give them some parts and they can assemble 100 PCUs per day. You will need additional people to test and certify, but you have a production line that is already making money.

In the avionics repair shop, it is a different ball game. The Air Data Computer (ADC) arrives with a complaint: "Intermittent airspeed and altitude flag in view." The six sigma black belt will ask, "What is an ADC? What is intermittent? What is airspeed? What is altitude? What is a flag? What is the meaning of flag in view?" When you try to explain these details, they will likely conclude that the technical knowledge is not important. No details please!

The problem: The unit is in the shop. The solution: Repair the unit. Putting it this way oversimplifies the process. You start with a pre-test and check the database for past removals of the subject unit. Then you conclude that it is a home sick unit and that CMM repair is not enough. If you cannot reproduce the defect, you must shake and bake it, test it, make it cold and hot and test it again, disassemble and do a microscopic check of all PCBs, and check all wiring, pins, and sockets for good connections or corrosion. You then visually check for mechanical damage, and, after you assemble it, test it again. If you test it on the ATE, you could repeatedly test it many times. A few hours or a few days later, the defect is found. The six sigma process flow, however, is broken. Instead of repairing three ADCs per week, you managed to fix just one. You are happy that the defect has been confirmed, that the ADC will fly again for many hours, and that the same serial number will not bother you for some time. The six sigma black belt will be annoyed by your shop's "poor performance" and will blame you for spending too much time on one unit. Six sigma planning is disrupted, and the normal probability plot shows extreme deviations. It will take weeks to level it out.

Management could decide to hire an external bureau because they are spending millions of dollars on six sigma experiments with no measurable benefit. The external bureau will start the whole circus all over again. They will use the presentations from their previous projects in an environment different than the avionics shop to impress your management. They will promise savings, increased productivity, and other great things. They will provide the list of LRUs, removals, and hundreds of parameters. They will provide graphs and fancy presentations. The conclusion will be that *IF* ... if you do x, y, and z, you can save X million USD in the next six months.

In theory it all works out. Just tell management what they want to hear, and they will be sold. The external bureau will sit aside and comment. They will say they are not knowledgeable about all the avionics details (the six sigma black belt would say the same thing). They know that the people on the work floor know how to perform better, but they don't listen because they want better equipment, and that costs cash: It's not a money-saving option, so it's not a good option.

The floor techs want to make repairs. They do not want to fill out additional forms or move Post-it notes along the walls. The people on the floor just want a simple production control system that doesn't disturb their activities.

What is the secret to saving millions? It's no secret. In the end, the responsibility for productivity and cost will be on the shoulders of the avionics engineers and technicians. They will be blamed for the failures, and their successes will rarely be noticed. Occasionally, they manage to save money and their jobs—not because they suddenly performed better, but because the six sigma staff changed the formula to measure the techs' performance for a better outcome. Six sigma projects should and *must* be successful. They are *always* successful. They *never* fail. Don't dare to say that a six sigma project was a failure. That is like swearing in church. The six sigma senior master black belt (Jedi master) is always right, and he never makes a mistake.

The above description is not a fictional example. It is based on my experience of the first few years after the introduction of the six sigma philosophy in my avionics workshop. Our projects were not generating income, and we were seen as failures. The six sigma approach was too theoretical to work in the avionics workshop, but nobody dared to intervene because it was dangerous to your career. I personally had to report to my manager several times because I made a critical comment about six sigma. In those meetings, I had to promise that I would shut my mouth and not talk negatively about six sigma because my manager would get into trouble for having such critical personnel. I was a rebel and management didn't appreciate rebels.

The Toyota Example

What I am trying to say with my criticism of six sigma is: Be careful, and don't go to extremes. Be realistic. Six sigma means one failure in 3.5 million operations. From an engineering point of view, this is a theoretical number, and achieving it is utopia. But every six-sigma-minded person told me that I have a stupid point of view and that Toyota proves that the six-sigma model works. The six sigma black belts were pointing to Toyota and introducing Japanese terms (*kaizen, kan ban, sama jama, poka, kutlul,* etc.) to their productivity model. In their eyes, Toyota's example showed that if you keep the materials under control, there will be no waste, and everything will follow the six sigma plan. In other words, in all Toyota processes, there is no waste. All the subcontractors are under control, and all the engineers are producing top quality work. Design is also superb. The cars coming out of the factory are perfect, and Toyota is what perfect car production without failures looks like. Only one car in 3.5 million is allowed to have a failure.

But based on Toyota's experiences in 2010, we know that Toyota's six sigma philosophy actually led to a huge fiasco. Six sigma is not working at Toyota. How else can you explain that Toyota had to recall 10,000,000 (10 million) cars because of failures? We are talking about *10 million* failed cars. That is massive. Toyota was not even able to reach one lousy sigma (69% defects). Sorry, Toyota.

I am not going to dig into the root causes of Toyota's failures. I am surprised that all six sigma master black belts were silent after that Toyota incident. I have had six sigma, lean, and kaizen jammed down my throat as methods to cut jobs and reduce costs. Toyota was held up as the shining example. Well, does it work? Let's hope that Toyota does the recall and fixes the safety problems a bit more efficiently than they make cars. The six sigma master black belts told all engineers to follow Toyota's example, and now there is not one word about the Toyota recall. Well, six sigma (black, green, and other belts) is it so difficult to admit that you were wrong? Apparently it is.

The Human Factor

Let me explain why I think that six sigma is a poor model. First of all, the Toyota car is a complex machine, almost like an aircraft but unable to fly. Let me compare it with aerospace and the cause of fatalities in aviation. Between 1907 and 1950, most aircraft accidents and incidents were caused by failing material and poor engineering work. Little by little, we engineers fixed the material and engineering problems. The accidents and incidents are still happening, but today the cause of accidents and incidents in the majority of the cases has shifted to the human factor.

Every aerospace engineer, from structures to avionics, is aware of the human factor. We all know about the dirty dozen human errors, we all know about the Swiss cheese model, and we all know that we should be dedicated to covering the holes in the Swiss cheese.

The slice of cheese is a safeguard, and the holes are a failure in the safeguard. Because we are in a dynamic environment, the holes are constantly changing position and magnitude. This is the problem. We in aviation know that we should design our LRUs and aircraft in such a way that they will be foolproof. We know that we need a suitable design to use an appropriate procedure, a proper way of testing, BITE, failure modes, MSG3 analyses, periodical checks, procedures, test procedures, incident reporting, occurrence reporting, and so on. We work in a very regulated environment. We are designing the slices of cheese and fighting the holes—managing unpredictable holes. We know how to do it, and we also know that one day we will be surprised that an accident has happened. Afterward, we will work very hard to correct the problem and prevent such an accident from being repeated. This is why we are the safest transportation segment. It is also the reason we don't like those six sigma priests telling us what to do. They told Toyota what to do, and we can see how that worked out. If Toyota had investigated the problem at the very first moment it was reported, they wouldn't have had to recall 10 million cars. (Or maybe one of the six sigma secret rules was to keep it quiet and not admit to any failure.) Nevertheless, six sigma is causing human fatalities because the six sigma gurus refused to face reality, and insisted on the success of their program over anything else.

The other possibility is that the six sigma gurus were afraid to tell their COOs about the failure, which is a human factor issue. When they could no longer keep the problem a secret, the issue exploded. The new point of view is that one failure in 3.5 million is not possible to achieve because of human factors. This is the Marijan Jozic theory. You might call it Marijanology.

Lean Six Sigma

So far, I have been spitting at the six sigma philosophy, and now I'll look at it from a more positive angle. Five or six years after the introduction of six sigma, the black belt Jedi decided not to call it six sigma anymore but to shift to "Lean Six Sigma." The six sigma activities provided some valuable tools that will stick around for a while. Using some six sigma tools is not such a bad idea. We do not talk anymore about quality, because the six sigma philosophy is not able to reach the optimal number of one failure in 3.5 million, but we can talk about process control. Normally there is a process to repair the LRU. We know how to do it. We need material, test equipment, manuals, and people. If you introduce an LRU into the avionics shop process, the technician will repair it and deliver it according to the CMM. Today this is not enough. The LRU must be repaired quickly, because you can save money on spares. That is when Lean Manufacturing kicks in. *Lean* means that the standards process should be measured and analyzed. There are a lot of measurement methods. I will not go into detail but will remind you of the famous words of Lord Kelvin spoken in 1883:

"When you can measure what you are speaking about, and express it in numbers, you know something about it; but when you cannot measure it, when you cannot express it in numbers, your knowledge is of a meager and unsatisfactory kind: It may be the beginning of knowledge, but you have scarcely, in your thoughts, advanced to the stage of science, whatever the matter may be."

This was back in the 1880s, and Kelvin was talking about measurement. Clearly, the six sigma or lean theory about measurement is not new. Remember that! It is something that engineers have known for at least 200 years, but it is now structured and branded by lean gurus. Your black belts can only get credit for structuring it, not for inventing it. Measuring the process is the key. Let's be clear: The measurement is the base. Only if you know the numbers will you be able to analyze the process and design the changes to improve it. Every engineer understands this. It is what we engineers do every day. We eat and breathe measurements and numbers. Establish the parameters to measure. Measure! Find the deviations and attack the biggest numbers. In a lean environment, it is the same process. Measure, analyze, and improve! After that, measure again, because you must determine the effect (if any) of your change. Repeat measurement day in and day out, and make it a standard practice. This will improve the integrity of your data. As an engineer, I am dedicated to this type of activity. It is feasible, it works, and it is immensely satisfying to see that your changes are having the desired effect. Engineering-wise, this is excellent stuff. At a certain point, the process is under control, there is stable flow, and you have created an optimal working environment.

The process will reach its limit, and you will not be able to optimize it further. You cannot make the process shorter or cheaper indefinitely. Even if your managers ask for it, it will not be possible. If they demand further optimization, they will spoil the progress you have made. You should maintain the optimal level and keep an eye on it. This status is called STAB OP (Stable Operation). If your managers say, "We want TAT [Turn Around Time] of 1 hour and the costs 90% lower," this would be a ridiculous request. Requirements should always be realistic and feasible. Setting an unrealistic target is foolishly short-sighted. It will frustrate everyone, and the desired improvement will not manifest.

Lean six sigma arrived after five years into the phase with a human face. It is odd to see its broad acceptance in the workforce, because if you do it well and with a human face, it brings success. First, six sigma black belts were on the wrong path for aviation. They wanted to analyze, measure, and improve, but their plan was theoretical drivel. We, the engineers and technicians, were making jokes and laughing about their new theories. We frustrated them, but they had access to higher management, and to prove that six sigma was working, they celebrated success after success, which later became failure after failure. Their approach was wrong, and they created a gap between themselves and skeptical engineers. To silence the engineers (the disbelievers), they engaged in deceptive tactics, and they made the expression, "You are either with us or against us" into a corporate policy. But, believe me, there are no engineers against analyses, measurement, and improvement. Engineers breathe analyses, eat measurements, and burp improvements! That is how they operate. But, "I know best," and "you should listen to me" never did and never will work with engineers. Why? Because engineers do the work, and they know best. As if both sides have the same polarity, engineers and six sigma black belts can only repel each other—just like the same magnetic poles will never draw each other, only repel.

Six sigma in maintenance is not the same as six sigma intended for production processes. The logical question is, "What is actually working?" It took me five years to ascertain the answer. At one point, I met a black belt who was not in the clouds with all the theory. The theorists were focusing on statistics, data crunching, calculations of the "voice of the customer," and more of the theoretical stuff. They said that they were listening to what the technicians were saying, but they were not.

Black Belt with a Human Face

This new six sigma philosopher was different. He joked about theoretical six sigma, but he himself was very pragmatic. He organized a kaizen event, talked to the techs, and within a few days, he was accepted, and the technicians and engineers were cooperating instead of opposing his involvement. The philosopher analyzed the situation together with the techs. He designed a spaghetti chart of the department, walked through the process, and, again, together with technicians, identified the waste. After that they decided (together) what to do to eliminate the problems. In a few days, the plan for the redesign was completed. It was obvious that the department was changing and had improved just by moving furniture and repositioning the test sets; introducing a few new items; improving lighting; scrapping a lot of old cupboards and unused test sets; and adding a few computers, printers, and other minor items. One example of efficient repositioning of furniture saved the team 15 hours per week (the complicated spaghetti chart was no longer complicated), which saved 750 work hours per year.

Using your own shop's hourly rate, you can calculate the financial savings. The money was enough to finance the replacement of furniture, improve lights, and install six more computers. Eliminating waste is called the *leaning* of the department. The people working in the department were happier, and happy people produce in greater quality and quantity. This philosopher gave a human face to lean six sigma. He was able to speak the language of the floor and help technicians work more efficiently. The team was no longer frustrated by waiting for computer access, waiting for materials, or having problems working in low light. The acceptance was 100%, because the technicians were able to see results in just a few weeks. Productivity increased, and savings in TAT was almost instant. This is what I call six sigma with a human face!

This is why I have an aversion to those black belts who show up and start to sell the theory but are unable to see the big picture—or the details—of the situation. Human nature is to work as efficiently as possible. We created a large and complicated process, and when the technicians are forced to fill out a lot of forms and make unnecessary changes, we can't expect that they will be cooperative and happy. If you help them create a pleasant, clean, light, and lean working area, they will endorse it and give you back the quality, efficiency, and quantity you desire. That is the human face of six sigma.

The list below notes a few effects of leaning the shop by a black belt with a human face:

- The shop will fix all backlogs and determine whether they have significantly more time to do more work. The shop was not able to manage the six sigma triangle. The work was coming in and there was too much waste. Therefore, the TAT was out of control. Now they are asking the development department to develop more capabilities because they need more work.

- They are also complaining to the sales department that they are not selling enough because not enough work is arriving in the shop.

- The inventory department will have an overstock of some LRUs because, due to the improved and stable TAT, there is no need for so many LRUs in the pipeline.

- The stores will complain that they have no storage space. Do you remember all the stuff that was in the shop waiting to be repaired? It is now repaired, and it is waiting on the shelf in the stores.

- The neighbor departments will start to complain to management that they also want a kaizen event and lean activity. They want more computers, a lighter workspace, and repositioning of furniture and test sets. They also want to be efficient, and they want to eliminate waste.

- The whole department is more peaceful. They do their work and have time to fix the most difficult problems. They even have time to think about the possibility of doing a proposal to engineering to accomplish repairs of piece parts that were, for many years, nonrepairable. This is saving a lot of money.

- The department has more time to solve no-fault found problems in LRUs, which have been giving them difficulty for the last few years.

The effect of the six sigma philosopher with a human face is amazing. In this instance, the image of the six sigma as theoretical and statistical drivel is turned upside down, 180 degrees. The technicians and engineers want to be lean. They want to be efficient, and they want to prove that they are skilled workers. Isn't that wonderful? The same techs who were exhausted from an excess of overtime hours are repairing more LRUs per month than ever before, without working any overtime. The whole department is happier. And happy staff produce quality *and* quantity.

Now we are at the point where some of our managers need a lean six sigma lesson. Until now, they pretended to understand the lean principles. But some of them are proving to be bean counters who don't see the big picture. I will admit that the whole leaning process has up-front costs. But don't forget about the large pile of cash that the lean shop is now bringing in. The shop is cleaned up and lean. The staff is proud of it, and management brings tour after tour to show off their new baby: The upgraded and efficient shop.

When management hears that another shop wants a kaizen event and wants to be leaned, they complain that people are using six sigma lean principles to get money to redecorate their shops. Management is missing the big picture again. There is another group of people in opposition: the bean counters. Their bean counting inevitably slows progress. They don't understand that you sometimes must spend money to make money. Below is an example of such limited thinking.

I heard a story about a professor at a university in Delft (The Netherlands) who calculated that if an operator flying a 25 wide-body aircraft improved the efficiency of their operation by 1% by eliminating waste in their processes, they would reduce costs and decrease spending by 35M USD per year. He was convinced that the bean counters would disapprove of his idea to invest 20M USD in order to get greater efficiency, though this would result in 15M USD returning to the bean counters as profit. This may just be a story. Technical people obviously have an image problem. The ecosystem is complex, and no one trusts the numbers. The same bean counters will easily approve an investment in new curtains in the cabin of the wide-body fleet (for 1M USD) without discussion. This is just the way it is. For now, enjoy the positive developments of leaning and embrace six sigma with a human face.

19

Human Factors

At the end of the 1990s, the authorities decided to mandate a human factors course for all airline employees. We didn't even know what "human factors" meant. A training center staff member was invited to our management team meeting to explain the term and tell us more about human factors. He explained that under certain circumstances there is a possibility that a mistake can be made by a person which can cause an accident, incident, or costly damage to materials or people. The trainer told us that every person working at our company would initially attend a one-day training session, and, after that, there would be a refresher training session every two years. This was obviously a serious development.

A few weeks later, our director mentioned in a staff meeting that the company didn't have enough instructors and was looking for people to teach the human factors course. Unfortunately, only a small number of staff had volunteered for the job. I mentioned that I believed this to be a fine ancillary position, and I was surprised there was not much interest. The director's response was, "I believe that you will be a good candidate for that instructor's job. You are hired." And he provided my name to the training department.

Back in the 1980s, I taught subjects such as Digital Technique and Aircraft Instruments, and I enjoyed it, but the rules were different in the 1990s. The training department required me to attend a didactics course before I could teach the new human factors course to staff. The only place I could take the training was at Dutch Railways. They had the course available, and they put me in a class with train drivers. It was fun to attend the training with this group. They had the same problem as my company: there were not enough instructors

© 2023 SAE International

215

to train their people about the new regulation and signaling in train stations, and therefore, they sent 20 staff members to the didactics course. Three of us from aerospace joined their class. After 25 hours in the class, we took an exam, and that is how I got my certification in didactics. This training also qualified me to become an instructor at night school and earn a few additional bucks. When life gives you lemons, make lemonade! For KLM, I still had to attend a three-day human factors course and an additional one-day training in order to lead that one-day human factors course for employees.

I became a human factors instructor at KLM and spent a half-year teaching the one-day course every week. I also trained myself to recognize situations related to human factors.

Coffee, Coffee, Everywhere

I'd like to explain human factors with an example. There is an incredible but true story about the removal of coffeemakers on short-haul flights back in the 1990s. It started with a flight attendant's complaint: The coffee from the coffeemaker is leaking everywhere except in the coffee pot, and we cannot serve coffee.

The technician replaced the coffeemaker and solved the problem. Cleaners cleaned the galley and the floor and made everything shine again. The next day, we had the same complaint. Another technician solved the problem by replacing the coffeemaker again. We had the same complaint again the next day. After a few days, the first technician was on duty again. He fixed the problem again, but he found it strange and went to the computer. He pulled the maintenance record and found that the last time he was on duty, the complaint was from a different aircraft tail number. He noticed that there were more coffeemaker problems on different tail numbers: five complaints on three tail numbers. There had been two aircraft with the same problem in the previous five days. On the rest of the fleet of 28 aircraft, there were no complaints.

The tech decided to shoot an email to his friend in the engineering shop noting the problem with the coffeemakers and asking him to check it out. It appeared to be a fleet problem. The engineer went to the repair shop the next day. It sounded fishy. In the previous six months, there were no problems with the coffeemakers, and now, suddenly, we had five in a row.

The shop tech was puzzled, too. He called engineering and complained that cabin attendants nowadays can't even use a coffee-maker correctly. He was getting his coffeemakers back in good shape technically, but they were dirty and soaked with coffee. They were removed from the aircraft, put into a plastic bag and sent to the shop with all that filthy fluid. They all were NFF, but he had to disassemble every coffeemaker to clean it. On the coffeemakers he had received for repair, the coffee collection plate was turned upside down. The coffee collection plate is a square-shaped plate with edges a half-inch high (Figure 19.1). On one side of the plate, there is an opening, and the coffee was supposed to be guided from the filter bag through the opening into the coffee pot. If the plate is turned upside down, the edges will be facing down, and the coffee will not be guided through the hole into the pot, but instead, it will leak out everywhere. The coffeemaker would then be one dirty, soaked machine. Because of the leak and all the coffee grounds, the best solution was to remove the unit from the aircraft and have it cleaned in the shop.

FIGURE 19.1 The coffeemaker's plate can "easily" be turned upside down, causing a coffee leak.

© SAE International.

© SAE International.

Coffeemaker number six arrived at the shop with the same complaint. It was dirty, and the coffee-collection plate was upside down. The engineer decided to notify the operations and requested that they issue a cabin crew bulletin telling the cabin crew to read the instructions before using the coffeemaker. The bulletin was issued two days later, which was very fast, but in the meantime, another two coffeemakers were removed from aircraft. This brought the total to eight. The next day, all flight attendants were informed, and the coffeemaker problems stopped. Bravo!

Two days later, the problem started up again, and coffeemaker number nine was removed. What is going on, and who is turning those plates upside down? The engineer checked the computer system again two days later and found another complaint. So far, he knew that the complaints had nothing to do with the quality of coffeemakers delivered by the shop. The complaints also had nothing to do with the aircraft tail numbers, because the problem was occurring on random tail numbers. It seemed strange that the complaint was written on the inbound flight but never on the outbound flight. It was also odd that every complaint was written during the inbound flight from Birmingham (UK). Could this be a useful lead? Probably, yes, because inbound flights from other cities were coming in without complaints. Was Birmingham the key to the solution?

The engineer decided to go to the gate before the next flight to Birmingham and talk to the cabin crew. He had already talked to the shop technician and the line maintenance technician, and it seemed that they were not the cause of this headache. He had also announced the problem to operations, and he decided that now was the time to talk to the cabin crew personally and see if they were causing the problem. In the meantime, coffeemaker number ten was removed.

The engineer was in the galley when the flight attendant arrived. He explained the issue and asked her to make him a cup of coffee, which she did. Everything went smoothly, and it was a good coffee. Everybody was happy.

The engineer left the aircraft, and the flight departed for Birmingham. Five and half hours later, the aircraft returned.

The complaint: The coffee from the coffeemaker is leaking every-where except in the pot. Coffeemaker number 11 was dirty and soaked. The plate was turned upside down. And the flight attendant said: "I didn't do it."

The poor engineer decided to travel with the aircraft to Birmingham the next day to guard the coffeemaker. Nobody was supposed to touch it without consulting the engineer. Somebody was turning those plates upside down, and he wanted to figure out who. Somebody was either playing games or making the same error systematically. After 2 hours of flying, they landed. The passengers disembarked. The crew disembarked. The engineer took the jump seat and started the watch. The new, fresh crew would arrive in 30 minutes. In the meantime, the aircraft would be cleaned and fueled. The cleaning crew arrived to clean the aircraft. The garbage was removed, and they started cleaning. The engineer was sitting in the jump seat monitoring the activity and waiting for the next crew. He wanted to see if the new crew was causing the problem.

One of the cleaners started working in the galley and began cleaning the coffeemaker. He was doing his job very thoroughly—and then it happened. I had a dog, and his name was BINGO. He removed the plate, cleaned it, and then reinserted it upside down! He was the person making the same mistake every time and causing the coffeemakers to leak everywhere except in the pot. He had no idea of the mystery he had created. He wasn't even aware that he was making a mistake.

The strange chain of events was caused by one weak (human) link. The cleaner was not aware that he was reinstalling the plates upside down. He pulled the plate out with his left hand, cleaned it, and reinserted it with his right hand. During that operation, he rotated the plate 180 degrees. The flight attendants never noticed whether the installation was right or wrong. They assumed that everything was fine, put the coffee in the filter bag, pressed the button, and carried on with other duties. When the leak became evident, they had to switch off the coffeemaker and hurry up to the aft galley to make coffee using the other coffeemaker. The flight was crowded, and there was not enough time to make coffee for everyone. The same cleaner always cleaned the forward galley, making the same mistake every time. It turned out that there were two days without problems. Why? The cleaner who was making the error had two days off, and on those days, somebody else was cleaning the forward galley (and not reversing the plate).

One unusual problem was solved.

What were the costs of that non-quality issue?

Removal and installation of coffeemaker:

- 0.5 hours × 11 coffeemakers × 50 USD man-hour = 275 USD
- Eleven shop visits with NFF 300 USD flat rate × 11 = 3,300 USD
- Cabin crew bulletin (writing, printing, and distribution) = 2,000 USD
- Engineering effort 30 hours × 100 USD = 3,000 USD
- Cleaning of galley and floor approximately 0.5 hours × 50 USD × 11 = 275 USD
- Total cost: 275 + 3,300 + 2,000 + 3,000 + 275 = 8,850 USD

Add to this customer dissatisfaction for those unhappy passengers who didn't get coffee or had to wait much longer for coffee.

Did the person cleaning the airplane realize that he had initiated this cost? No, and we cannot really blame him. Perhaps he wasn't instructed on how to clean the coffeemaker and reassemble it after cleaning.

From an engineering point of view, the real cause of this problem is a poor design. It should not be possible to reinsert the coffee collection plate upside down and cause such coffee leakage. If the design was better, so that the plate could only be inserted in one way, the above complaint would not have been possible.

Therefore, my fellow engineers, be aware that every design you launch should be foolproof, easily accessible, and easy to maintain. Check it out carefully before mass production. You could be blamed for the cost of future problems.

This is an example of human factors. One of the dirty dozen is likely to show up, because since the beginning of time, we humans have made errors. Making mistakes is human, but we must be aware of them if we are to reduce and prevent them and not purposely—or accidentally—create more.

The poem "The Rime of the Ancient Mariner" by Samuel Taylor Coleridge is about an ocean voyage on a ship. When I think about this coffeemaker problem, I can't help but recite the poem, replacing the word *water* with *coffee*:

Coffee, coffee, everywhere
And all the boards did shrink.
Coffee, coffee, everywhere
Nor any drop to drink.

Dirty Dozen

For those who are unfamiliar with the dirty dozen human factors, a summary:

1. Lack of Communication: This is simply the failure to exchange information. Training should focus on not only how this happened but also what safety net will prevent it in the future. Very simply, in good communication, "the mental pictures must match."

2. Complacency: This happens when we become so self-satisfied that we lose awareness of danger. It is sometimes caused by overconfidence, which creeps in as we become more proficient at what we do. Awareness of this insidious contributing factor is one of the safety nets that helps to reduce it.

3. Lack of Knowledge: With constantly changing technology, this contributor to error is more common than we might think. Add to this the fact that the average human only

retains about 20% of what they learn, unless they use it often. Training is one of the best safety nets we must have to help avoid this human error.

4. Distraction: This is anything that takes your mind off the job at hand, even for an instant. Our minds work much quicker than our hands, and thus, we are always thinking ahead. Any distraction can cause us to think we are further ahead than we are. This contributing factor is known to be responsible for at least 15% of all aviation accidents.

5. Lack of Teamwork: The larger an organization becomes, the more common this contributing factor is. Because teamwork is constantly evolving and changing, it must be constantly worked on to prevent accidents from occurring. Teamwork is difficult to create and very easy to lose.

6. Fatigue: This is considered the number one contributor to human error. It is insidious, and, generally, the person fails to realize just how much his/her judgment is impaired until it's too late. Fatigue seldom works alone but is a contributor to one or more of the other dirty dozen.

7. Lack of Resources: The lack of resources to safely carry out a task has caused many fatal accidents. For example, an aircraft is dispatched without a functioning system that normally would not be a problem and then encounters circumstances in which it becomes a major problem.

8. Pressure: Pressure to be on time is ever-present in the aviation industry. We are very time sensitive and many decisions center around that fact. More than 64% of pressure-caused errors are caused by self-pressure. One must learn how to recognize and deal with pressure.

9. Lack of Assertiveness: The lack of assertiveness in failing to speak up when things don't seem right has resulted in many fatal accidents. However, assertiveness also calls for listening to the views of others before making a decision. Assertiveness is the middle ground between being passive and being aggressive.

10. Stress: Stress is the subconscious response to the demands placed upon a person. We all have some stress in our lives, and it is not a problem until it becomes excessive, and we have distress. We must learn how to manage stress, or it will manage us with a high probability that human error will occur.

11. Lack of Awareness: Lack of awareness occurs when there is deficient alertness and vigilance in observing. This usually occurs with very experienced persons who fail to reason out possible consequences to what may normally be a good practice. One of the safety nets for lack of awareness is to ask more "what ifs" if there is conflicting information or things don't seem quite right.

12. Norms: The word *norms* refers to something that is "normal," or the way things are typically done in an organization. Norms are unwritten rules followed or tolerated by the majority of a group. Negative norms detract from an established safety standard.

Importance of Human Factors

Aviation is approximately 100 years old. We learned a lot in those first 100 years. We figured out that in the beginning, there was a lot of focus on solving technical issues and problems. Great engineering power was required to build a sound aircraft. Previously, when engineers decided that they had created a good aircraft design, they were stunned when a bad accident occurred. After the accident investigation, there was proposed improvement, and the engineers were able to design an aircraft or an aircraft system that would not cause the same type of accident again. After many years, we were able to design a near-perfect aircraft. Just look at the statistics: In the last few decennia, few accidents were caused by technical problems. Most were caused by human factors.

The awareness of and the importance of human factors is essential. Most people involved in aviation today have had human factors training and are familiar with human factors issues.

We know about the dirty dozen, which cause most of the errors. We are aware that human factors and dirty dozen issues are serious threats to safety, and the cause of the majority of aircraft accidents. If you are aware that anyone can make a mistake and that many factors can contribute to making a mistake, then you are well on your way to improving safety and quality. Safety starts with you. Be aware of safety nets and do your part in contributing to safety.

Some avionics shops have a remarkable history. They have been in business for many years, and there are a lot of excellent techs who have been maintaining LRUs for many years. These experienced techs are difficult to replace. They do their jobs and repair the same aircraft parts for 20 years. They grew into their professional roles with LRUs, and they contributed to the improvement of those LRUs. The new techs on the job got an OJT from the older techs, and after some time working, they are up to speed. They can repair the LRU and deliver quality. Occasionally, QA staff shows up in the avionics shop to check how the techs are doing their job. QA just wants to confirm that the avionics techs follow procedures. Infrequently, they find that the experienced shop tech cuts some corners and does not respect the rules of aviation. Such disregard for procedure can cause one of the dirty dozen human factors. Now we'll review a few other well-known human factor events.

Toyota Recalls

At Toyota, they knew about problems with their vehicles for a long time, but by the time the chain of events unfolded, it was too late to deal with them in a simple and cost-effective way. In 2009–2010, the problem spiraled out of control. Toyota had to recall millions of vehicles, and they became an industry laughing stock. Some found it satisfying that they failed, because we had enough of Toyota as an example of the invincible six sigma principles.

This attitude is also not OK, but it shows the frustration of those in the industry.

The Toyota case may repeat itself in other organizations if staff feel pressured by unreasonable demands for productivity and cost savings. Of course, you should lean your organization, but don't be unreasonable, because if you are, the fate of Toyota may await you.

The Toyota case is an example of human factors due to the following from the dirty dozen:

1. Lack of Communication
8. Pressure
9. Lack of Assertiveness

A similar situation occurred with the Space Shuttle Challenger and the failure of those rubber sealants. Again the combination:

1. Lack of Communication
8. Pressure
9. Lack of Assertiveness

In reviewing the above, I gained a lot of insight into human factors. It became my professional default, because I was always looking for reasons why people react and do what they do, not only in my workshop but also outside it. This knowledge helped me solve numerous nonconformities in the shop as well as understand human nature and why people act the ways they do.

Schipol Shuttle Bus

I still become angry when I think about the behavior of flight attendants on the airport shuttle bus. Picture it: There is a shuttle bus between the main airport building and the crew parking lot at Schiphol-Amsterdam Airport. The bus is just an ordinary public bus without a compartment for luggage. There is not even a shelf for luggage on the bus (Figure 19.2).

Eric Gevaert/Shutterstock.com.

The crew, when coming from a long flight, is supposed to put their bags and suitcases in the rear part of the bus. Sometimes there are people on the bus and the crew just puts their suitcases in the aisle or by the door. Sometimes, the door area is so filled with suitcases and other luggage that no one can leave or enter the bus. It is a short trip, and the chance of an accident or other emergency is remote, but still possible. It is ground transportation, and nobody is concerned about this possibility. As an experiment, I started a discussion on the bus with the flight attendants, the owners of the luggage. My observation was that if somebody put their luggage that way in the airplane, every one of them would go into action and warn the passenger about the unsafe situation. The passenger would be forced to remove their suitcase from the aisle.

Now, on the bus, they all ignored safety and blocked the doors and aisles (and they all had flight safety training). They were wearing their official uniforms and were representing their companies. Of course, I was the bad guy (the engineer) who verbally attacked the cabin crew, pointing out what safety really is. I am just a stupid engineer who was talking too much and bothering them. But I am worried. I am worried that there is a chance of meeting those flight attendants on a flight. They didn't get the message. They considered their training as part of a show and think they are actors who repeat the same show on every flight. They haven't grasped *why* there is a regulation to store all luggage away and keep all emergency exits accessible. They have learned to warn the passenger if that passenger's bag is placed in the corridor or by the emergency door. So, they know better, but they are doing the same with their bags on the bus. If you offer a friendly warning about this, they become incensed. They are blocking the doors with their luggage and making other people unsafe. They are not thinking that the bus might have a collision or catch on fire. All passengers should be able to evacuate the bus quickly, but they can't because of the luggage blocking the exits.

The same crew who rules the airplane behave like a pack of sheep on the bus, not thinking and just following along. Safety does not stop at the aircraft door. Safety does not stop at the end of your workday. Safety is always a requirement. Every one of us has just one life to live. Why waste it by behaving stupidly? Why expose yourself to an unintentionally dangerous situation? Train yourself to recognize unsafe situations. There is always a way that safety can be improved, and if you can, you should improve it. Safety does not stop at the end of your working day.

This bus example was a combination of Complacency, Fatigue, and Lack of Awareness. And now you might want to know what I did to improve this situation. It was simple. My neighbor is a flight safety instructor at ground school, training department, so I asked him to emphasize in his flight safety training the importance of recognizing that everything that they learn in the training, they should carry over into real life, on buses, in hotels, at airports, and at home. Don't obstruct the walkway.

20

Repairs Not Defined

With the introduction of 737-NG, A320, A350, 777 and 787, intellectual property (IP) became a factor in aviation. Before those aircraft entered the scene, nobody in aviation talked about IP. You could call a supplier and ask for the CMM or a drawing, and they would readily send you the package via FedEx, no problem at all.

Airlines changed the situation when MD-11 entered the aerospace. Honeywell visited KLM and offered a special deal for maintenance of components for the whole Honeywell line of products installed in MD-11. Our head of engineering was enticed by the deal. All of his colleague managers proclaimed that we should take the deal. I and my fellow engineers could not oppose them; our boss made such decisions unilaterally. A similar situation occurred at other airlines. The bottom line was that OEMs contracted a lot of aftermarket work. Previously, they only sold the new components and piece parts. Now, they started to sell service repairs. Around this time, I was in Phoenix for a course at Honeywell. The instructor complained that they were getting an extreme amount of repair work. His sales group didn't think about that when they sold the contracts: They did not have capacity in their shops to do the contracted work.

Then something unexpected and unique happened. They began to realize that there was a lot of money to be made in the aftermarket. The OEMs increased their capacity, hired new people, and started to serve customers by doing repairs. That was the new business model. Airlines that had traditionally repaired components started losing their position in the aftermarket. Some larger airlines established MROs and continued to fight for jobs. Others decided to outsource all repairs and invest in logistics departments to manage all the outsourcing.

This was the second part of the 1990s, and engineers noticed that OEMs were delivering more and more reliable LRUs. In the 1980s, engineers would be happy if an LRU was doing up to 10,000 hours MTBUR. The LRU in modern aircraft at the end of the 1990s was doing 20–25,000 hours MTBUR. The effect was that the number of shop visits was much lower. In order to cover department costs, the MROs started to contract third-party work. Test sets were expensive, and shops needed decent workflow to recover the investment.

The introduction of 787, 737-NG, 777, A350, and A320s brought some oddities to the shop. The test sets were extremely expensive, and reliability of LRUs was way higher than 50,000 hours MTBUR. The suppliers implemented IP. Airline MRO (based on a PSA [Product Support Agreement]) was able to repair its own components, but for third-party work, they had to sign an IP agreement with the OEM and pay royalties. Suddenly, they had to share the pie with the OEM.

Stories about the shrinkage of CMMs popped up, saying that OEMs removed repair procedures, IPCs, and a lot of data from CMMs. Everyone believed it. Under pressure from my managers and the purchasing department, I had to check 2,500 CMMs and indicate which (if any) OEMs were removing chapters from CMMs. After a lot of work by my whole team, I did not isolate a single CMM with chapters deleted. Nothing, nada! There were a few cases where chapters were reshuffled and instances where the IPC was removed and issued as a separate manual or where the CMM was very thin, but every PCB was handled in a separate CMM. All that effort and panic for nothing. We did notice that many CMMs included very little or no description or procedure for repairs. Sometimes, at our request, the OEM would insert the repair procedure.

To lower costs, MROs had to work smarter. The MROs wanted to introduce the PMA parts and repairs. This was sometimes a big problem—not a technical problem but a legal problem. For example, say you are running the shop and your sales department signs a contract for pool support of a customer. Most customers leased the aircraft and signed a contract requiring that no PMA parts be used in the aircraft. If you were to dig into the contract, you would likely find that the PMA parts for the engine and primary structure are subject to that clause in the contract. The simple PMAs used in LRUs were not mentioned, but sales staff don't understand that; often, they were former car salespeople who had changed jobs to sell aviation pool contracts. I don't blame them for not understanding parts and repairs; I blame them for not listening to the engineers who do.

Eventually, engineers lost the PMA battle. It is near impossible to use the PMA in the pool component. In my shop, we have isolated many components with internal PMA parts. Officially, we are supposed to remove those parts and install the original part; otherwise, the LRU is not CMM compliant. MROs do this, and people complain about it. But if you send them a drawing authorizing that PMA part, they will keep it in the unit.

CRT Display

Repair of an LRU with repair not described in CMM is a different story. One of the best-known cases was the display of a 747-400. Rockwell Collins announced in SB 2001-10 that Cathode-Ray Tube (CRT) displays would become obsolete and unprocurable in mid-2003. Of course, they would support it for some time, but the end of production would be in 2003. They offered an attractive retrofit to LCDs, which were drop-in replacements. But if you had to replace six displays per aircraft you would run into high costs. In that uncertain time, a small company from Australia entered the market offering the repair. Their plan was to keep the old CRT displays and repair them by replacing the CRT tube. This company provided an exchange of CRT assembly for a fraction of the Rockwell price.

I had difficulty with my QA department and my management, because everybody was extremely fearful of starting that type of repair. Eventually, based on the Australian CAA certificate and

FAA 8110 and EASA approval, we were in business. For the next 20 years, we sent broken CRTs to Australia for repair. At the beginning of the millennium, I found out that the Australian company was using Toshiba tubes. Toshiba delivered tubes to the OEM and to that Australian MRO. My engineers purchased one tube just to experiment. It was a purchase of a few thousand dollars, and my boss was angry at me for spending money on my hobby projects. When the tube arrived, my engineers went to DORA and started reverse engineering. After a few weeks, they showed me their results: The display with the Toshiba tube was working perfectly. We collected all the data and made some drawings and descriptions, and I sent them to my favorite DER, Bob. Bob was a repair specification DER, and after a few weeks and some additional documents, we received FAA approval for the repair. It was 15 years late, but still, we were now able to do what the Australians were doing. I knew that if I played it right, I could save millions on CRT repairs. But my inexperienced superior completely blew it. He decided to go to purchasing to renegotiate the extension of the contract with the Australians. He took the box with the refurbished CRT to the negotiating table and said, "If you don't drop the price, we will do it on our own," and he showed them the refurbished Toshiba CRT.

The Australian stopped the negotiation and left. I learned later that he departed for Tokyo the same day. There he negotiated an exclusive contract with Toshiba. I received an email from Toshiba that it would be impossible for them to sell us additional tubes. This backfire was the result of our own stupidity. Instead of saving millions, we saved nothing, but nobody was blamed for the failure.

Ducommun Dimmer

Another repair story regards the Ducommun dimmer in the 777. Unfortunately, this is another example of stupidity, but at least this time we managed to save some money. In a 777 crew rest area, dimmers were installed for dimming the reading lights (Figure 20.1). This was nothing special, just a simple dimmer. You could switch the light on and off and dim it. The dimmer was never electrically defective, but the plastic button was of inferior quality, and in 95% of cases, this was the cause of removal. It was considered the consumable part, and every time the dimmer was broken, they replaced it and threw the broken dimmer away. We had a small room where one tech used to collect scraps and tried to find out if a repair was possible. He called me in and showed me a dimmer. He had at least 50 dimmers in a box. I found out that there was no CMM for the dimmer and the price of a dimmer was 2,000 USD. We used to order 20 dimmers per year, meaning a cost of 40,000 USD per year for that poor-quality, simple dimmer.

FIGURE 20.1   Crew rest light dimmer on the left, with expanded view on the right.

 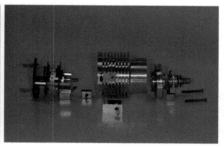

© SAE International.

© SAE International.

I had a few flight attendants in my department, and I asked them if they knew about the dimmer in the crew rest area of the 777. "Oh yes, we know about it. We switch it off by hitting it with our slippers." At least I now knew one root cause of the defective plastic cap on the switch. Since there was no CMM, we couldn't get it in the shop to repair it. We would not be able to release it with a certificate (8130 or F1).

One day, the FAA came to visit us and do an audit. My management always selected me to deal with authorities. I had met the FAA rep previously, and we had a good relationship. I told him about my plan to issue the CMM for the dimmer, and he asked me to send him the CMM and supporting data, and he would see to getting it approved. My best engineer, Marcel de Lange, took the project. He fabricated the CMM, did some testing, and got the 8110 from our DER (Bob) with "Recommended for Approval." That all went to my FAA friend in Frankfurt, who forwarded it to FAA ACO in New York. In a couple of months, I received the FAA approval for the dimmer CMM. This was unique: For the first time in the long history of my company, we had our own approved CMM.

Our QA got nervous and did everything possible to prevent me from repairing the dimmer in our shop. All the arguments and data didn't help. I stopped arguing with QA and asked my friends from a partner MRO shop in Los Angeles if they would be able to repair the dimmer for me and provide the dual release, and of course, they could. So now I had a situation where a shop in LA was repairing our 777 dimmers using our CMM, and my own shop was not allowed to do it. At least those 50 dimmers (and many more) were repaired for a fraction of the price of a new dimmer, but it was still a frustrating outcome for my team.

I didn't want to place a sticker in the crew rest area with the text "Don't use slippers to switch off the light," but I went to my neighbor, who was a flight safety instructor, and asked him to warn the crew not to use slippers to switch off this light. And that was the end of the dimmer project.

PCU Keyboard Repair

The most legendary repair at KLM during my tenure was the repair of the PCU installed in the business class seat. I was passing by the office of one of the cabin engineers, and he called me in to show me the PCU. The OEM of the business class seats was a Japanese company. They were cheating with certification data, and the FAA decided to suspend their operation (and license). Therefore, they were not delivering parts for seat repairs. The whole wide-body fleet had those Japanese seats installed, and we had two years to replace them, because they were not compliant with requirements. In the meantime, we were allowed to repair mechanical parts by fabricating them in our facility. Among those vulnerable parts was the PCU keyboard. This was just a simple plastic keyboard with a few membrane switches. There was a switch for the reading light, a switch for the flight attendant call, a switch to start the massage, and switches to move the seat to different positions. Finally, there was a switch to bring the seat to the upright position for landing and take-off (Figure 20.2). The problem with the keyboard was that sometimes the switch for bringing the seat to an upright position was not working. Because of this, the operator had to deactivate the seat. There was already one airplane flying with a deactivated business class seat, meaning that no revenue was generated from that seat. The OEM was not delivering parts, as all the parts were already spent.

FIGURE 20.2 Inside view of the keyboard used to control each business class seat.

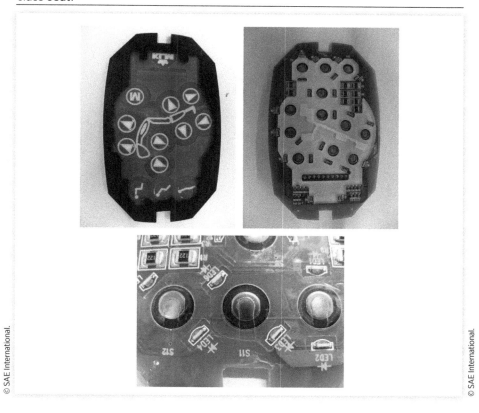

The cabin engineer had about 150 broken keyboards in a box. He offered it to me and asked if I could do something with it. On the way back to my office with the box, I met the head of the engineering department, who expressed concern about the keyboards and asked me to do anything I could to fix them. After that, the purchasing rep told me that they had purchased all available keyboards and that there was no hope of fixing the problem. I knew that this would be a difficult project, but I thought it worth a try. I also knew that I would have to destroy some keyboards before we could draw any conclusions.

After some experimenting, I managed to peel off the plastic keyboard cover. The first thing I saw was a lot of white material accumulated under the membrane switches and everywhere under the keyboard. It looked like a dry-cleaning solvent. After a few telephone calls, I discovered the following. After the flight and when all passengers have disembarked, a cleaning crew enters the aircraft to clean the interior. They spray the plastic parts with cleaning solution and dry each surface with a dry cloth. It looked like a tiny quantity of liquid penetrated through the small holes under the keyboard and contaminated the PCB below the plastic keyboard. When the solvent dried, it isolated the contacts, and the keyboard stopped working.

Back in DORA, we managed to clean the surface, and the keyboard started to work. There were two problems: The first was to peel the keyboard without damage, and the second was to attach the keyboard back to its original position. The second problem was easily managed if the first step was executed properly. Luckily, I had a variable warming plate, which could heat the keyboard to a specific temperature. At that temperature, the glue became liquid, and it was easy to detach the keyboard. We destroyed a few keyboards before we established that the best result was at a temperature of 72°C. The problem was solved. After that, we sorted out the paperwork, got the EASA approval from our engineering department, and we were in business. Or at least, I thought so.

Our local EASA inspector came in to see what was going on. Why? Someone obviously reported me for suspected dangerous activity. It was quite a disappointment that somebody reported me to the authorities without coming to me first and asking what was up. I didn't have anything to hide. The inspector checked the paperwork and witnessed the repair, and said: "This is excellent and a good job! Everything is as it's supposed to be. Congrats." It was success with a bitter aftertaste. I never learned who reported me, but I realized that I had to watch my back. I secured the sales of the business class seats, but found little pleasure in doing something good for the company. I was challenging the status quo, running after projects, standing firm behind my engineers, and pioneering, but due to jealousy (and fear), some people were only interested in criticizing me and shutting my work down. This was disappointing.

As the Latin saying goes: *Per aspera ad astra*; "Through hardships to the stars." When I look at my life, I see that my greatest periods of personal growth came after particularly challenging moments or experiences outside my comfort zone. Every time I managed to overcome disappointment and find something to learn from the situation. I have saved the company hundreds of thousands of dollars, and today, nobody even remembers the success of my actions. One of my fellow engineers, Mr. H., once said, "A good engineer cancels himself!" We called this the First law of Avionics Engineering, and it is exactly the situation I found myself in time after time. If you do your work well, everything runs smoothly, and nobody even knows that you are the reason everything is running so smoothly. Because of that invisibility, it is easy to make yourself redundant. If you are a lousy engineer, you always get help, and everybody knows about your projects because you always need to escalate them. People might get the impression that your projects are difficult and that you are important. This is usually a false impression, but this is how it goes.

21

787 Entered the Aerospace

The first 787 was officially delivered to All Nippon Airways on September 25, 2011. A new aviation era started with this delivery. I was working at a major European airline that was flying the 787, and my focus in this chapter will be on that bird and my related experiences.

The introduction of a new aircraft is not easy, especially when the last introduction was 20 years earlier. In those 20 years the world of aviation—and the world—had changed. The people who had introduced the 747-400 were on the edge of retirement, and their experience introducing an aircraft vanished as they left the industry. So the next generation had to invent the wheel all over again. In addition, the whole aviation environment had changed, and there was much more attention on IP. At the AMC, we had a number of seminars about IP, data sharing, data ownership, and so on. Suppliers were no longer willing to share data. We, the airlines, and suppliers used to be like friendly coworkers who worked together to fix problems, but now we were more like adversaries.

Engineers within airlines were not informed that the rules of the game had changed. What happened? Airframers teamed up with suppliers. Suppliers were paying a lot of development costs for the new airplane, and they wanted to recover those costs on the aftermarket. Therefore, the PSA (Product Support Agreement) had to be renegotiated. This is an agreement between the airframer and the suppliers. When an airline purchases an airplane, it gains the rights to obtain all documentation, but only for the maintenance of its own fleet. Here was the problem. The equipment installed in the new aircraft was extremely reliable, so the airline operating fleet of, say 25 787 aircraft would not have a large flow of components through the shop. At the same time, test equipment was made extremely expensive, and no data was made available to enable techs in the MRO shop to build it on their own.

Another issue was that airframers wanted to reduce the number of suppliers involved. Therefore, they forced the suppliers to work together. That introduced finger-pointing and avoidance of responsibility. You know the game: One supplier says, "That component is not ours but theirs." The other party would say, "We understand the problem, but according to the contract, we are not allowed to give you the documentation." I created a so-called Agony List with 24 of these problems. The list was discussed weekly with the airframer, and one by one a solution was created for each problem. This was the real agony: those lengthy and complicated telephone conversations.

SCEA

At the airframers' suggestion, the AMC committee gathered a group of engineers to work on the SCEA (Standard for Cost Effective Acquisition). These issues were extremely sensitive, especially with lawyers. I always said, "Lawyers are the kings and queens of the company." But when they interfere with engineers, everybody loses. The lawyers were panicked that we were calling the SCEA document a *standard*. One corporate lawyer attended a meeting to hear and see what we were doing. Then he opposed the fact that we called it a *standard*. Finally, he decided that we could call it a standard because it was an above-board document. As I mentioned earlier, there was a PSA between the airframer and the suppliers. This document also defined the rights of the operator and the operator's MRO shop. Everything outside the PSA was open to negotiation between the MRO and supplier. This was an area of third-party work and "forbidden" for the aircraft operator. As I said, the aircraft operator's workshop could only work on components installed in its own fleet. The function of the SCEA document was to cover that gap, help MROs play the game fairly, and also make it possible to do third-party work.

I was not only involved in writing the SCEA, I also was the chairman of the SCEA group. Being the chairman and attending all the meetings made me an expert in this area, which was useful. We also defined examples of delegation letters in the SCEA. This is the letter that an operator is supposed to use when they decide that some p/ns will not be insourced but outsourced. The operator delegates their rights to get the data to the MRO. This method of data sharing was acceptable to the airframer and OEMs (suppliers). Our third-party customer should send such a letter to the supplier, but our third-party customer was not aware of this new procedure. Within my MRO, I requested that the sales team update the customer on this new condition. Their response: "No, we can't do that, our customer will be angry if we ask that. We are at the point of closing the deal. Are you crazy?" They refused to inform our customers that they would be required to send a delegation letter to the OEM. They signed the contract, which couldn't be unsigned. Sales eventually realized that I was right (it took them 8 months), and the delegation letter became a standard procedure.

The procedure started to roll out with the delivery of the new aircraft. A few maintenance contracts were signed, and life became easier. Airframers delivered a list of parts installed in the aircraft long before the aircraft was delivered. That list provided a good overview of parts, costs, reliability, and so on. The list was used for the distribution of capabilities in our workshops. That process took us 12 months. During that time, we acquired a wealth of knowledge about the new aircraft.

New 787 Technology

The 777 had just a few fiber optic cables. The new aircraft, 787, have many fiber optic cables. Also, the data bus used is not the slow ARINC 429 data bus, but the much faster ARINC 825. My engineers had to learn these modern technologies. There was a lot of new technology in the 787. One day somebody told me that the

ARINC 825 bus was also installed in cars, including the Mercedes brand. It is called the CAN bus. Because I didn't have an instructor available at my ground school for the ARINC 825 bus, I decided to hire someone from the auto industry. The instructor happily came to our facility, stayed for a few days, and provided a one-day course for each of my 70 techs and 20 engineers. That was well-spent money.

Another new technology was the Ball Grid Array (BGA), which was used for the first time in airplanes. In the 787 there are no classic CBs. We used these for 75 years in civil and military aviation, but the 787 has electronic CBs. The pilot could see on his screen which CB popped out. There was no ceiling panel with CBs in the cockpit, but there was a cart file cabinet in the aircraft with PCBs emulating CBs. In the 787 there are 175 PCBs that emulate CBs. Microchips on those boards were soldered with the special BGA technique. This type of soldering was a highly technical skill, and you needed a special course, microscopes, and even an X-ray machine to check the soldering. I organized that training for my shop as well. Even staff from the flight simulator department wanted to attend the course. I immediately made a deal. The course was free, but they had to give me a one-hour session in the 787 flight simulator. This attractive deal was sealed, and that is how I was able to do three or four take-offs and landings in the 787 simulator. It can be fun to be an engineer. You sometimes have the opportunity to play with some awesome and expensive toys.

My engineers were finishing the 737-NG capabilities development. We had just a few more capabilities to go, and we would be at 80% of the flow. Some of the engineers were already switching to 787 capability development. Selecting capabilities was not that difficult. Let me rephrase this statement: Making a To-Do List of capabilities is not difficult. The list is a function of the reliability of the component, price of the test set, the relationship with the OEM, and the price of external repair. A solid list could be compiled without difficulty. We had learned a lot from the 737-NG project, and that knowledge could be used for 787 capabilities with a few considerations. One important and logical item is how many of each LRUs were installed in the aircraft. We were used to having two or three or a maximum of four LRUs of the same p/n in the aircraft. In the 787 there were 175 PCBs as electronic CBs. That was a high priority. Furthermore, at every window, there was something new. There was a push button to dim the window, and passengers loved to play with it. Each of those 120 windows has a dimmer. Such items are ideal for capabilities developed in the avionics shop. Can you imagine the flow when the fully mature fleet of, let's say 30, aircraft is flying? It is a great deal of work for technicians in the shop.

Anecdotes

Now for two interesting anecdotes from my time dealing with 787 capabilities. Around that time, I attended the AMC, and one evening I met a remarkable gentleman. His name was Neal Speranzo, and he was one of the top managers of a well-known LRU manufacturer from Arizona, Honeywell. Neal was probably the most important person at that AMC for us operators. We were drinking tequila in a hospitality suite and talking about aviation. We both were already aviation dinosaurs—or at least, that is how Neal described us. When I was leaving the hospitality suite, he told me to contact him if there was ever anything that he could help me with. I said, "Yes, I will do that," but I knew that I wouldn't. I was just an engineer at a major European airline, and he managed 25,000 people at an OEM. There would always be somebody in a lower position that I could bother. It seemed that Neal could read my mind. "MJ," he said. "Don't worry. Your email will not stay on my computer longer than 10 minutes. Just do it if you think it is necessary." "OK," I nodded, and this time I meant it.

Several months later, Honeywell was giving me resistance on providing the lists of parts that I needed for initial provisioning. That initial provisioning was the list of parts that are failing more often than other parts. I told them that their competitor had provided me with a great list. The content of the list was:

In one year, we had 125 shop visits of that part number, and the following parts were used for repair.

Here is the list:

- Part a. Qty 6
- Part b. Qty 24
- Part c. Qty 17
- Part d. Qty 1
- Part e. Qty 1

[And so on.]

From such a list I can extrapolate the quantities required for use in my shop. But Honeywell's EU representative didn't want to cooperate. I contacted Honeywell in the US to no effect. Just this response: "We don't have such data." I didn't believe this, and I contacted Honeywell reps in France, the UK, and the US, but still received no cooperation at all: We don't have any data about p/ns used. It was time to send an email to my friend Neal, my last hope. I was not sure that he would even remember our conversation of six months earlier.

> "Dear Neal,"
> Yada, yada, yada ... and
> Please help me.
> MJ

In less than an hour, Neal called me. This was far beyond my expectations—I was hoping for an email response with a suggested contact. He indicated that he understood my problem and was going to push some buttons. Nothing happened during the next 24 hours. Then he called me again. "Did somebody contact you?" and I said, "No." "OK, please standby," he said. "You will be called back." And then his staff started to call me offering help. The next day I had my list.

I did not include this story to amuse you. This story is about a top manager within a major OEM who cared about his customer enough to become personally involved in meeting the customer's needs. This is the true spirit of aviation. Neal and I met again at the AMC in 2022, this time not drinking tequila but red wine (Figure 21.1). We revisited this story, which had happened 15 years earlier, and we both still remembered the details. Now we are

FIGURE 21.1 At the AMC conference with Neal in Memphis, 2022, 15 years later.

dinosaurs, but dinosaurs with a lot of aviation experience. What a great person is my friend Neal Speranzo.

HF Transmitter

I have one more 787 story; this one is about the HF transmitter. The HF transmitter was like a transmitter we had in another aircraft type. Therefore, the capability was established quite early in the process. One day, Erling Brom, one of my best technicians, came to me with a story about a strange fluid in the HF transmitter. That fluid entered the unit from the top side. In the top cover there were ventilation holes, making it easy for any fluid to penetrate the unit. On the top side, just below the cover, is the power supply. The power supply was contaminated and burned due to multiple shorts.

FIGURE 21.2 Erling Brom adjusting the HF transceiver of 787 aircraft.

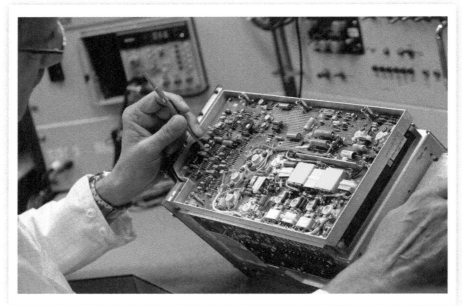

Courtesy of Marijan Jozic.

Erling was complaining that it was the second unit with this problem he had received this month (Figure 21.2). I decided to keep an eye on those HF units.

We agreed that Erling would warn me every time such a unit came to his desk. Over time, we noticed that we were receiving units from external customers as well as units from our own fleet. There was significant damage to each unit, and the OEM refused the warranty claims. I decided to submit the question to the AMC and introduce the problem in an open forum discussion.

At the AMC several months later, we engaged in a heated discussion on the HF transmitter problem. Based on my question, other airlines checked their fleet and found out they had the same problem. It was a 787 fleet problem. Repair costs were significant, and airlines were not happy. The OEM was diverting all claims and complaints to the airframer. The reason for this was simple: The airframer specified the HF transmitter with ventilation holes on the top of the unit. The OEM had built the unit according to specification, and they took no responsibility for any defect caused by the specification. The airframer told us that an SB was pending to fix the problem. We went home and waited patiently.

My technician Erling kept complaining. He had collected 50 units with the problem, and his warranty claims were consistently rejected. One year later, we heard the same story at the AMC. The OEM diverted questions to the airframer, and the airframer spoke on the success of the aircraft SB. But nobody wanted to hear about the collateral damage, which was the burnt power supply boards and water intrusion. When the pressure got too high, everything melted, including the HF case (Figure 21.3). Finally, the airframer accepted the warranty claims. I was happy Erling was so thorough in notifying us of the repair details for every unit. He also saved photos of the damage, and there was no further discussion with the airframer, which simply awarded the claims. Over the course of three years, this HF transmitter flaw cost a significant amount of money, but at least this story had a happy ending.

FIGURE 21.3 Damage in the HF transceiver caused by water intrusion.

Courtesy of Marijan Jozic.

It is unfortunate that it took three years to resolve a problem that should have been fixed in a year. This example shows that pointing fingers is counterproductive. It was not productive in handling my Agony List, and it was not productive in the HF case. With just a modicum of cooperation, the industry could have saved millions in USD on repairs and avoided the frustration endured by Erling, me, my engineers, warranty managers, and many others. I wish that there were more airline execs like Neal Speranzo. The aviation world would be a better place.

22

Reorganization (Again)

What makes corporate reorganizations so popular? My first encounter with reorganization was at the end of the 1980s. Within my company, it was known as the dinosaurs reorganization. They told us we should adapt to the new situation; otherwise, we would become extinct like dinosaurs. After numerous meetings and many hours of discussion, it was decided to merge some engineering groups and introduce a planning system for engineering jobs. Afterward, there were months of discussions and almost no significant change in the organization. Then it was decided to decentralize the engineering group. Before the reorganization, we all were in one building doing our jobs. After reorganization, engineering was spread throughout the maintenance department, and the consequence was that engineering became larger and more complicated. Eventually, we all kept doing the work we had done before, with some added staff to make it more practicable. All the meetings and discussions and moving of departments were a financial drain, but the reorganization was declared a major success.

Honestly, the only useful outcome of the reorganization from my perspective was the work-hour planning system for engineers. This was an excellent tool because we all learned how much time we spent on our activities. That was a big improvement.

FRIFRA Jobs

The system worked as follows: Every task was given a job number. Managers were able to monitor the use of work-hours, and they were able to assign a certain number of work-hours to a specific job.

One day our boss decided to hire a young admin to monitor us engineers via the planning system. The admin's name was Frifra. If our boss assigned a job number to an engineer with the precondition that it would be more than 50 work-hours and had a firm deadline, Frifra pulled the job number from the system. That type of job was called a FRIFRA job by the engineers. In the monthly report to management, there was now a separate chapter called FRIFRA jobs. It was a good system, and within three months, everybody was used to it. After three months, Frifra's contract expired, and he left the company. But the name FRIFRA stayed. Even 20 years later, we still had FRIFRA jobs in the system. If somebody asks about the origin of the job name FRIFRA, hardly anyone knows the answer. Just a few of us old-timers can explain that Frifra was a guy from Morocco who worked at KLM for a few months, and he used to assign the jobs. Nobody believed it.

I have spent time in the project office, which was part of the maintenance department. Every four years, there was a reorganization. It seemed that every time there was a new boss, he reorganized the office completely. Those were difficult times. The workforce learned what to do and how to behave in a new work model, then a new boss arrived, and the department was turned upside down again. Such messy and uncertain times are a killer for production and efficiency. In addition, staff are unhappy under these circumstances. And unhappy people are not really motivated, so productivity declines.

One of the reorganizations led to the establishment of self-supporting teams. There were no managers, and the team had to do all the management tasks. Managers became coaches. Teams didn't know what to do, and coaches didn't know how and whom to coach. The idea was to lower the number of (expensive) manager positions. Eventually, the newly designed coach function was renamed team representative. After two years, the concept was reevaluated, and we returned to a normal situation with managers and staff. Approximately 20 years later, new managers (who were hired straight from universities into their first management jobs) decided to establish self-supporting teams. If you tried explaining that we attempted that system 20 years ago and it was a failure, they immediately put a sign on your back with the name "trouble-maker." And guess what? One year after they established self-supporting teams, they had to revert and assign a manager to each team.

Avionics Reorganization

After my transfer to the avionics shop, we got a new manager. The reorganization was announced, and I was part of the reorganization team. The group of 15 people on this team had to design the new company unit. I learned a useful process on this team. First, you define all jobs and processes in the department. This could be done in an Excel sheet with a few hundred-line items. Then, you cluster the processes, and finally you assign the functions. This is just a few top-level steps, but it can save months of lead time. Of course, there were numerous meetings and discussions. Once the new setup was ready, it was necessary to run a few test cases to see if everything would work. If that was satisfactory, we had to compile the workplace instructions and descriptions of each new job function. The final step is to inform all staff and then assign the job functions. Once everything is done, and the switch is flipped, all staff have to learn how to do their new jobs. This time in the shop, everything was fully transparent, and honestly, the success was obvious. The organization was more flexible and very lean. All communication lines were fast and open, and people were quite happy and settled after a couple of months under the new system. A certain amount of time is required for people to get used to any new situation. Everybody noticed the change was positive, and it worked well. That reorganization was settled and running for six years.

You're Fired

Then a new boss came to the avionics shop. It was OK in the beginning, but then he decided that we should reorganize. This time no one from the shop was assigned to the reorganization team.

He designed the new setup with an external party and some internal gurus from another department. It was quite an extensive reorganization, involving many departments.

In the end, the boss called the management team to a meeting room and told us that we were all fired. That was unexpected! You are doing the best possible work for the company, business is good, and then we were all fired. We were told that we could keep doing our jobs until a newly hired person took it over. The next announcement was that there would be a list of positions published on the intranet, and we could apply for any job on the list. My job was on the list. I could apply for my job, which I did. But there was another job that got my attention. It was similar to my current job (Capability Development Manager). The new job was in the sales department, and it was called Industry Development Leader. The position involved the same duties I currently performed but helping our customers.

I made an appointment to interview for my own job. I went to the interview with my boss, and he told me that I was not qualified for my job. He said that I underestimated the job function, and it appeared that I thought the work was easy, but it was actually difficult. He wanted someone with a master's degree in engineering science for the position, and I "only" had a bachelor's degree. My 35 years of engineering experience were not considered as qualification enough. I left his office and took the stairs to go down to my office. My mobile phone started to ring, and on the line was the manager in charge of the sales department. Without an in-person meeting, he offered me the job in the sales department as Industry Development Leader, and I accepted. When I returned to my office, which was near my department, the whole group came to me asking if I would stay in the department and be their boss after the reorganization. When I said no, everyone was disappointed. I was sorry to leave my dream team after 14 years.

My new job paid better, which was a plus, but my former department would have a new boss. I decided to give each of my engineers a bottle of wine and invite them to a restaurant for a cocktail party. They gave me a present and wished me good luck. One week later, I packed up my stuff. In the morning, all the fired managers met our successors, shook hands, and wished them good luck. My chapter in the avionics shop was closed.

My successor left the department after three months. After that, the shop engineering department had no manager for three months, and then they got another academic in the role. He managed the department for six months and then he left, too. Then they spent another seven months without a manager. The department was degraded due to that stupid reorganization.

In my final year as department manager, we launched 24 capabilities for 787 LRUs. My team was operating superbly. They were cooperative and happy, and happy people produce quality work. It felt like anything was possible for the team—the sky was the limit. In the 12 years I was there, we evolved from a museum to a top-notch avionics shop. Was that reorganization necessary? Absolutely not! It was just the brain fart of a new high-level manager who wanted to leave his mark on the company. It took a lot of time and effort to create a cohesive, productive organization, and they destroyed it in a few weeks. The second year after I left the shop, not one new capability was developed. I asked myself how that was possible when everyone had stayed on the team (except me).

I have seen many reorganizations, but that one was the worst. The workshops had 400 people and a turnover of a few hundred million USD, and you just remove all managers in one day and install people who have never been in an avionics shop? Some of them had never heard of LRUs, SRUs, ATEs, job carts, and FARs. This was a recipe for disaster. There was no handover and no training period of any kind. Was this a smart way to manage a transition? I don't think so. But it was not my business anymore. I had to keep calm and carry on.

It was not the first time that I had left a department, and the whole department collapsed afterward. A high-ranked manager once told me: "You, MJ, are very dangerous for the organization. Too many people and too many processes are dependent on you. If you win the lottery and leave the company, our department is lost." That was quite a compliment to me but also a bad sign for the organization.

He was probably right. I had a bad track record. When I left avionics systems engineering, the department collapsed and never recovered. They are at the edge of existence, doing insipid SB evaluations and no projects at all. The business project office was shut down two years after I left. D checks were outsourced to China and are not done anymore. And now the engineering shop was barely surviving. Perhaps it is a coincidence, but it always happened after I left and took my experience and institutional knowledge with me to another group.

Repairs Development

A spin-off in the latest reorganization was a new department called repairs development. The department was filled with people who had never developed one single repair. The reason for such a department was that sales negotiated contracts for third-party work and accepted the clause that we would only use original parts. That clause killed a big portion of PMAs, because costs could not be lowered with the use of PMA parts. Additionally, DER repairs were not acceptable. Those two items made repair development very difficult—actually, impossible.

Now, there was a department that would do the impossible and teach the engineers a lesson. The academics in the department (I have nothing against academics) made a long list of repairs, something my engineers had done in the past. Then they designed fantastic presentations for top management showing how efficient and productive they were. Obviously, they had plenty of time to fabricate great PowerPoint presentations with flying sheets and flashing numbers. I had made a similar presentation the previous year, and nobody paid attention. My Excel sheet was simple because I didn't have time to spend hours adding bells and whistles to a presentation. They had the time, and, of course, their presentation was impressive.

After the repairs development team had bragged about that data, they had to do something new. For their first project, they talked to my friends from the department servicing aircraft evacuation slides and learned that in the slides there is a small battery pack.

Those battery packs had to be replaced every two years. There was a big box of old battery packs containing plastic tubes full of penlight batteries and a switch. When the slides are fired up, the switch should flip and provide the power for the light. Those tubes (battery packs) were a few hundred bucks each. But because we had a lot of old battery packs, they decided to repair them by changing the batteries. Because they were not engineers, they went to the aircraft manufacturer and asked for the SB for the repair of those inexpensive battery packs (tubes). The aircraft manufacturer issued the SB, and then the repairs development team came to my engineers and asked for a calculation of the costs of that repair. If the repair required two work-hours or more, it would not be worth doing. The costs were calculated to require about 2 hours per repair (cutting the tube open, replacing the batteries, putting the batteries in the tube and welding them in series, closing and sealing the tube, and finally the paperwork). This was obviously on the cusp of breaking even, based on their calculations. But all non-recurring hours of managers, trips to the airframer, discussions with the airframer, and costs of SB were not considered. The option to go to the OEM and renegotiate the price of the battery pack also was not considered. That option is not a repair development, so it was useless to them.

If you are even a bit technical, you will understand that this was a ridiculous project. Eventually, the whole issue was forgotten. Reorganizing and creating a department with a goal of saving millions on repairs was not constructive. Rather than establishing a new department, the shop engineering teams could have been bolstered by one or two engineers so that all engineers could spend some time in repair development. The sales department could also be instructed not to close any deal that would limit the shop activities. But, again, this was not relevant to me anymore.

New Beginning

Usually, when you get a new job, someone provides you with training or at least orients you to the department/position. This person might serve as a resource as you learn the new job.

This time, there was no one to serve this role for me. It was actually the fifth or sixth time in my career that I started a new job and there was no one to orient me. I had to change jobs three times due to reorganizations. Each time I managed to move to a position that paid better than the previous one; in that respect, the reorganizations were beneficial for me. Oddly, in 40 years at the airline, I was always rejected when I applied for a job in another department that was advertised on the intranet. And each time I was offered a job without any formal application. I am not afraid of reorganizations. If you do your job well and cooperate during reorganization, the payoff can be immense. Just don't take the change personally, and everything will be all right. Change is good, and more money is always welcome.

FIGURE 22.1 My last day at KLM, handing over the torch to my son and receiving a farewell award from my dream team.

23

Global Operation

One day I started in a new department called Customer Services. I was one of a kind. All the other team members were customer support managers or sales managers or customer support officers, but I was a capability development consultant. I was the only engineer in the department. After my first day, I was reminded of this saying by the philosopher Erasmus Rotterdamus: "In the land of the blind, the one-eyed man is king." Paraphrased, this means, "If surrounded by people less capable or able, someone who would not normally be considered special can shine." I hoped to shine in my new position, but I was feeling overwhelmed with tasks, analyses, and business trips. The last thing that I wanted was to travel. I had been perfectly happy with four trips per year: three meetings and the AMC. Now I was burdened with customer visits. Some of our customers were far away, and I didn't like having to travel every other month to Indonesia, China, Morocco, the US, or France. In this new ecosystem, technical knowledge was not very important. Politics and manipulation were more essential skills. In order to survive, I had to learn how to operate in this new environment.

My primary focus was customer support. The sales team negotiated the contract, and often, to get the customer to sign the contract, they promised to help the customer establish capabilities. The noble idea was that the customer outsourced the maintenance of LRUs to our shop, and once the customer, with my help, established the capabilities for a certain p/n, that p/n would be removed from the contract and maintained by the customer. This was a great idea, and many customers bought it. But it took ages to establish the capabilities at the customer's location, which was to our advantage. Sales used me to buy time and keep the customer happy, and the customer would send the components to our shop for repairs. I knew that it was likely a mission impossible for the customer to develop capabilities, but I dedicated my time and knowledge to helping the customers move toward this goal.

Morocco

One of the customers was in Morocco. Therefore, I had to do two-day trips to Casablanca every two or three months. The job was hopeless, because the customer had no desire to establish the capabilities. They had closed all their shops and outsourced the components. The technical staff were laid off, retired, or assigned to other jobs. Even with great determination, it was impossible to start working on capabilities, because there was no money to relaunch the shop. I could only review the situation provided in the report and advise the customer on what to do. Nothing changed. Their expensive test sets (ATEs) had been switched off for five or six years. They pulled the plug and did nothing after that. In the meantime, all backup batteries were empty, and all the software was gone. The software saved on their hard disk was OK but outdated. Updated versions were required. My best advice was to sell the unused equipment and make a few bucks. But that was impossible because the government had paid for the equipment, and it would be seen as corruption if somebody sold equipment purchased by the government.

At least I have seen Casablanca. I even found Rick's Café, a restaurant from the movie *Casablanca* starring Humphrey Bogart. I also discovered that Rick's Café was fake. The entire movie was filmed in Hollywood, and the street scenes were filmed in Paris. The famous words, "Play it again, Sam!" were never even spoken in the movie. Ingrid Bergman says, "Play it, Sam. Play *As Time Goes By*." Well, I had to make the trips to Casablanca, so I investigated and came to the disappointing conclusion that they make things up in Hollywood!

Short trips like this were generally pleasant. On every trip, we had a few hours of downtime to go to the local market and practice negotiating. I had completed a Karas course on effective negotiations, and the local market was the ideal place to practice. They liked haggling, and I had fun. I was not buying big things or spending large amounts of money, but it was not important what you bought; it was important to go through the negotiation process and get the best price. Once I purchased a baseball hat, another time sunglasses, then a fake designer watch. The guy selling fake designer watches told us, "Original copy. No chocolate inside!" We joked about those embellished sales statements for many years afterward.

Asia

I took numerous trips to Asia, where I supported my friends in avionics shops in Shanghai and Indonesia. My Chinese friends from Shanghai were charming. They welcomed me as a friend, and it was gratifying to help them. I felt fortunate to have the opportunity to see the shop and meet the techs and the staff. We had an enjoyable time together, and after that experience, my support from Europe went more smoothly. We all knew better how to collaborate.

I had a sizable network, and if there was a need for something, I often was able to provide assistance. On one occasion, the Chinese team called to ask if I could help them acquire materials. They could not buy the materials they needed because they had no contact with a particular company and were unsuccessful getting a response from the company. I found the company's representative in Germany, and he connected me to their Asian sales office in Singapore. The Singapore office provided information for the sales representative in Shanghai. It took me a couple of calls to get the correct person. A few days later, my Chinese friend called to say that the representative of the company they had been looking for was in his office, and he was very thankful for my support. I was overcome with feelings of fulfillment.

Kenya

One of the sales managers invited me to go with him to Kenya (Figure 23.1). There was a customer who wanted to organize more local capabilities, but they didn't know how. So, we went with a small team to assess the situation and make a plan of approach. I considered the trip as just another chance to help a customer with capabilities. As usual, we started with a meeting and a presentation about capability development and the well-known 4Ms. Next, we had a tour of the facility, and then we went back to the meeting room for discussion and planning. The fleet was small, and the components were reliable, so there was not much to gain from capabilities. Quick wins were also not obvious. The only components that might be candidates for capabilities development are simple components and some hub support stuff. Regarding hub support, it was straightforward. They already did wheels and brakes, which can be a big hitter. Other items, including ELTs, battery shop, and headset repairs, were in place. We decided to fabricate a list of p/ns and their removal rate. I would analyze the list and come back with p/ns for which we could develop capabilities. My contact was a young engineer who was not very experienced but was willing to learn.

FIGURE 23.1 Me in Shanghai on the left, and in Nairobi on the right. They used to call me the traveling engineer.

Courtesy of Marijan Jozic.

The month after our first meeting, we had numerous email exchanges, and I taught the engineer how to do the calculations for a business case, and how to select the components for capability development. Eventually, we made a thorough and workable list. Unfortunately, after we had made progress and were ready to launch the projects, all contact via sales ceased, and the project was stopped. There would be no more travel to Kenya for me.

Indonesia

Often, I traveled to Indonesia for work. For people who are unfamiliar with the Indonesian aerospace situation, I would like to mention that Indonesia is a very big country with a huge domestic market. It takes around 3 hours to fly from end to end. Thousands of islands were connected through the air. Unfortunately, the trip from the EU to Indonesia is quite long, and on one occasion, I made a costly and time-consuming blunder. I didn't notice that my passport would expire in five months, and for Indonesia, visitor passports must be valid at least six months upon leaving the country. My itinerary was Amsterdam – Kuala Lumpur – Jakarta, and back. On the layover in Kuala Lumpur, everybody was required to leave the airplane and return in an hour. After a 14-hour flight, it felt good to walk through the terminal building. There was a tropical garden and a place to buy fresh juice. This was my ritual every time I came to Kuala Lumpur on the way to Jakarta.

Upon returning to the gate, I had to show my boarding pass and passport, and they stopped me. My passport was expiring too soon. They removed my bags from the aircraft and left me in Kuala Lumpur. I called my friends in Jakarta, and they told me that they would arrange everything with customs and the police, but I should try to come over on one of the next flights. There were three flights, but I was unable to make it onto any of them. I was not allowed into Malaysia for the same reason: six months passport requirement. The only solution was to intercept the return flight and go back to Amsterdam. Luckily, there was a seat for me on the return flight. So, I had a 14-hour flight from Amsterdam to Kuala Lumpur in economy accommodations, spent 7 hours waiting for the return flight, and then endured a 14-hour flight back to Amsterdam. Thirty-five hours after my departure, I was back in Amsterdam. That was a lot of flying in a short time and without accomplishing anything at all.

When I was able to return, cooperation with my Indonesian friends was beneficial. There were many young and knowledgeable engineers on the Indonesian team. We were working hard on capabilities and to define a scope for the contract about cooperation between our companies. My only concern was that my company would force me to go to Indonesia for two years as the project required. Many years earlier, I wanted to go to Wichita to work on a project, but my company didn't want me to go. This time they actually asked me to go to Indonesia. I didn't want to go and therefore they sent another team member. He did a lot of work in preparation for the big contract about cooperation and I was supporting him from the home base in Amsterdam.

To mark the signing of the contract, there was a big celebration. I was, of course, invited, because I negotiated the contract and I had to run the whole project. By then, I had been to Indonesia many times and met much of the local staff. I considered the local engineers as my engineers, and once again, I had a good team. The celebration included the signing of the contract and a short ceremony with a few speeches by the top managers of both companies. After the ceremony, drinks were served. One local engineer asked if I would like to take a selfie with him at the podium. Why not? We stepped up to the podium and took the selfie. Then another person came over, and another, and another. In no time, a line of people was waiting to take a photo with me. I didn't know that I was so popular! I gladly stood there and said "cheese" for each picture. Then my VP came to the podium and spoiled our fun. "It looks like you are a local celebrity. But I believe that they want me too in the picture." That was arrogant of him, but he likely felt affronted, because soon the line dwindled away. At least I had my minute of fame.

For many months after that, I helped my team in Jakarta. They traveled in small groups to Amsterdam for OJT at our facility. Soon afterward, they got their certificate, which concluded the capability development for that p/n. We established tens of capabilities in that year. It was a strong team capable of doing excellent work. With just a little bit of steering, the team was operating superbly. On some of my trips to Indonesia, the local team took me to karaoke bars after work. In addition to capability development, I discovered that they also had excellent singing capabilities. The team was performing well, and we had fun. I was afraid there would be something or someone who would spoil the experience, which seemed to happen to me whenever my team was operating on the highest level. We had a cooperation contract. The team was operating smoothly, and we all had a lot of fun. But a little voice was telling me that the future was uncertain. Who knows what will happen?

24

COVID-19

The A330 was shaking in turbulent air. The bumpy ride woke me up above the Indian Ocean. There was turbulence every time on that stretch of the flight. That is the stretch between Sri Lanka and the Strait of Malacca. Normally, I would fly on a KLM flight to Kuala Lumpur and then on to Jakarta. That flight arrived in the evening at Soekarno-Hatta International Airport in Jakarta. This time I flew on a Garuda flight to Medan and then continued on a domestic flight to Jakarta. This used to be a 777 nonstop flight that arrived in the morning, which was a better option for me, because I could go to sleep for a few hours and then take a swim in the hotel pool to get going. For unknown reasons, Garuda decided to replace the 777 with an A330. This airplane was not able to fly a nonstop flight to Jakarta. We landed in Medan on Sumatra Island on February 4, 2020.

I was aware that there was a strange new illness in Asia that had originated in China. Nobody seemed to know any details about it, but it seemed perfectly safe to go on my trip. On Dutch TV they were saying that the illness was not contagious. Even the Dutch Minister of Health said that there was no danger for the Netherlands and Europe because there is no direct flight to Wuhan from those locations. Wuhan was thought to be where the virus originated. On February 4, 2020, I was in Medan waiting for my connecting flight. To my surprise, most of the people at the airport were wearing masks. There was even a group of Chinese flight attendants wearing masks and gloves. That was strange. Where I had flown in from, there was no sign of the illness, and here in Medan, everybody was masked. I decided to wash my hands and keep my distance from people as much as possible. A few hours later, in Jakarta, it was the same story at the airport and in the hotel.

Normally, when I visit Jakarta, I'd stay for two nights and have two days of meetings. It was no different this time. I was happy to go home. My flight back was via Kuala Lumpur, and I again saw many people wearing masks at the airport. Upon arrival in Amsterdam on the 7th of February, I noticed that things were business as usual in Europe: no masks, no precautions. I arrived home on a Friday and had the whole weekend to recover from jet lag. That evening I got a fever. I spent the next day in bed, still had a fever, drank a lot of tea, and slept. Sunday afternoon, I felt better, but my significant other had a fever. By Monday evening, she was better, but our youngest son had a fever. After two days everyone was OK, and within a week, everything was back to normal. Did I bring COVID-19 (Coronavirus disease of 2019) home with me from Indonesia? Nobody will ever know. At that time, COVID-19 tests were rare.

Exposure Risk

I had agreed with our partner airline in Indonesia that we could do some OJT (On the Job Training) for the VHF for three of their techs. They were supposed to come over at the end of February. But by then, people were talking much more about COVID-19. I told my shop techs that three people were coming from Indonesia for OJT for VHF. They immediately started to complain about the risk of exposure to the virus. I called the medical department and asked about COVID-19. The answer was, "Don't worry. That is not a very contagious illness. If you are sitting around the table, nothing will happen. It is very difficult to get it." I discussed it with my techs again, and they asked me to postpone the OJT. After some discussion, I decided to play it safe and listen to my techs. I went to management and explained that I wanted to postpone the course due to the risk of exposure of our techs to COVID-19. I didn't want to expose our techs, and I also thought it important to listen to the technicians and keep them happy.

Explaining this to my manager was like cursing in church. My direct manager was very angry with me. "We must respect the contract. No way, we cannot postpone the training. Tell your team that we will provide the OJT no matter what, and have them shut up and stop being babies. Corona is just a simple flu." I was becoming more and more convinced that we should postpone the OJT. I informed the partner airline in Jakarta that the training was under discussion and that there was a possibility that it would be postponed due to COVID-19. I also assured my techs that the training would be postponed, and they calmed down. Now I had to manage the situation with my managers and alleviate their concerns about the contract obligations. I knew their concerns were unfounded, because I negotiated the training during a trip to Jakarta, and I knew the date was not firm. Moreover, I had a solid relationship with my partners in Jakarta, and they would understand the situation; of that I had no doubt.

I needed a manager who would support me, and I found one. He would be in the staff meeting and promised to put the training postponement on the meeting agenda and support my position. After that staff meeting, my manager came to me and said, "We decided to postpone the meeting. We cannot expose our people to the risk." He completely turned 180 degrees from, "We should respect the contract" to, "We care about our technicians." After that encounter, I postponed the OJT indefinitely.

Grounded

The next day, the news hit Europe: President Donald Trump decided to stop all flights from the EU to the US. My management complained about losing money because the US fleet was grounded. When I said that Trump had made the right decision, I was told,

"Because of your Trump, you will lose your job." Three days later, management decided to send us home, and we were all supposed to telework. This was funny because before the pandemic, working from home was taboo. If you asked your manager if you could work from home for just one day, it was not acceptable. Now we were all sent home to telework indefinitely. I was sitting at home and answering mail. One week later, there was no mail. For me, the world stood still. That spring we had lovely weather, and I spent time in my garden sitting in a lounge chair, drinking coffee, and watching the grass grow. Sometimes I forced myself to walk a bit, but due to lockdown and curfews, it was not possible to walk very far.

The first of June was coming, and it was a big date for me. It was my anniversary: forty years employed at the airline. My 25th anniversary was lousy because my manager forgot about it. I was looking forward to getting that diamond pin for my 40th anniversary and having a blast of a celebration, but thanks to the pandemic, there was only a small ceremony. I was allowed to go to the office with my significant other and my two sons. The secretary handed my wife the diamond pin, and she pinned it on my jacket. It was too dangerous for my boss to approach me at a close distance because COVID-19 was a very contagious illness. This was the same guy who had said: "Coronavirus is just a simple flu!" My wife and my sons were in the same bubble as me, so there was no danger to them. There were six people at my celebration: My two sons, my wife, me, my boss, and the secretary. Not very exciting after 40 years in aviation. There was no need for speeches; just take a picture and go home. That was the last time I was in my office. That was the glorious day of my 40th anniversary in aviation: June 1, 2020.

I didn't like the pandemic situation because I was not in control. My future and the future of humanity were uncertain. It was impossible to make any decisions because nobody knew what would happen the next day. I had a lot of aviation knowledge, but there was nobody who would accept the transfer of my knowledge. I was not the only engineer in that situation. All my fellow engineers whom I had met at conferences or on business trips had the same problem. Aviation was looking for excuses to get rid of people. Transfer of knowledge was not a priority. Before COVID-19, aviation was an exciting career. During the pandemic, nothing was happening because the fleet was not flying. The only work available was to park the aircraft and do a simple check every six weeks. The airline industry, like many others, was hemorrhaging money.

At one point during the pandemic, a company from Italy contacted me. Somebody in the industry recommended me to them. I am the guy who makes things happen. They had designed an interesting filter unit. The filter was a wire grid made of tungsten, and around the grid are LEDs with visible light. When air flows through the grid, the combination of light from LEDs and tungsten (as a catalyst) kills all bacteria and viruses. They asked if I could certify their filter for use in aircraft. It could be installed in one package with a HEPA filter. The only requirement would be DC power and an indicator somewhere in the cabin indicating whether the LEDs were working. It was not difficult to design the installation because the wire grid did not stop the air flow at all. But there were two problems: One was to convince aviation, which was already in financial trouble, to do the modification and pay for it. The second problem was me. My company offered me a voluntary retirement package, and I decided that my days at the airline were coming to an end. I liked the idea of those filters and saving the world, but it was not to be my mission.

I accepted the voluntary retirement approximately four weeks later, and four months after that, I just had to return my laptop, safety shoes, and phone to KLM. The glory days of aviation were behind me. I say glory days because I am convinced that I actively participated in the best 40 years in aviation history (Figure 24.1). In those years (between 1980 and 2020), everything was possible. It was paradise for engineers. Many new aircraft types were introduced, many huge projects were implemented, and many adventures transpired. After 2020 the world will be different, but I will always be able to say that I was there when history was made. The glory days of aviation are over. When everyone returns to work following the pandemic, aviation will be a whole new ball game.

FIGURE 24.1 Marijan Jozic enjoying navigating through the golden years of aviation.

Courtesy of Marijan Jozic.

25

Trust that an Ending Is Followed by a New Beginning

Summer 2020. COVID-19 is everywhere. The whole world is turned upside down. The flying fleet is not flying. Airlines are losing money, and almost every country is on lockdown. The company sent us to work from home, but there was not much to do. Still, the summer of 2020 was perhaps the most beautiful summer I have ever experienced. The sky was crystal clear, perhaps because there was no flying and much less pollution from cars. One effect of this was that the weather apps on our phones became unreliable. The learning algorithm had to adjust to the new situation, and it was a couple of months before it could provide a consistent weather prediction again.

During the summer of 2020, many companies couldn't afford to pay salaries to people sitting at home doing next to nothing. KLM offered us a voluntary retirement package, and without much hesitation, I accepted. I considered myself to be at the top of my career (for the fourth or fifth time), but I decided to accept the package and go. It is better to leave when it's your decision than to wait and end up in a situation where somebody else would decide for me. I had a positive feeling about leaving after 40 years. My career had been an adventure, and not many engineers were so lucky as to work in almost every corner of the airline and maintenance business.

I started as a technician repairing little boxes of the entertainment system and then spent several years in the calibration lab. After that, I became a flight guidance engineer. For most engineers, the job in systems engineering was the end of the line. They work there until retirement. But a few of us did more. My next step was as a senior engineer and then project manager. That changed into program manager (which means doing multiple projects) and then modification manager. For a while, I was a business analyst, then I became the account manager for D and C checks. After several years, I made a complete change of environment and went from aircraft to components as head of the shop engineering department. I loved this job, and I did it for 15 years (Figure 25.1). The department was successful, which was not always recognized as a good thing by management. After a reorganization, I was transferred to the sales group and tasked with helping our customers develop capabilities and run their own shops. That was my whole tour through the engineering maintenance department.

FIGURE 25.1   In my office close to my dream team at the avionics shop engineering department.

Courtesy of Marijan Jozic.

In a parallel track to my workflow, I was active at AMC, which was managed by ARINC as part of Industry Activity. I started small and at a later date became the AMC Chairman. I was also Chairman of the Mechanical Maintenance Conference for one year, until those two institutions merged into the Aviation Maintenance Committee conference. I built my entire career in that separate lane. After 7 years as a Chairman, I stepped down in 2019. During those 20 years, I received the Volare award, Roger S. Goldberg Award, and Man of the Decade Award. Those were awards from the ARINC Industry Activity side. From my own airline, I was given the TQP (Total Quality Performance Award).

I also spent 10 years helping to organize Aviation Knowledge and Career Day at the University of Amsterdam, and I was nominated twice for the Aerospace Journalist of the Year award (2004 and 2005)—and I am not even a journalist!

After my long career, I often asked myself what I should do with all the experience and knowledge I had collected. Whether I like it or not, once I leave the airline, I will take a lot of knowledge with me. People can become obsolete. This is serious, because if knowledge is lost, the company must start from scratch when they need that information. If the company is smart, they will be proactive and take care that knowledge is preserved. But most companies do not plan ahead. Occasionally, some knowledge might be stored in a document(s) and saved on the server. If not, too bad—it's gone forever.

Three Knows

This is why I came up with the theory of the three *knows*. The first know is *Know-How*. Know-How can be learned to a certain extent. Most of this type of knowledge can be learned: self-study of CMM, AMM, and SRM; courses; and many books and drawings. Generally speaking, the engineer will be able to maintain and even modify aircraft and keep it flying for 40 years. Know-How is transferred from engineer to engineer, and all the details can be found in educational resources. This know, Know-How, is easy.

More important is the second know, *Know-Why*. Know-Why is usually only stored in the heads of people. Only those who Know-Why also know where this information is noted or recorded. The ability to find it is crucial. Let me give you two examples.

The first example is the time needed to evacuate the aircraft. We know that, in case of an emergency, the aircraft must be evacuated within 90 seconds. The question is: *Why* 90 seconds? Why not 60, or maybe 120? Based on the 90-second rule, we calculate how many doors you need in the fuselage. Should it be 10 or 8 or maybe 6? Try to find out why it was decided that 90 seconds should be the regulation! It is not that easy. I searched for the answer for many years. Then an engineer from Brazil told me that the 90-second rule was established because the average person can hold their breath for a maximum of 90 seconds. If there is smoke in the cabin, the passengers are supposed to hold their breath for 90 seconds and still be able to leave the airplane. Ninety seconds (1 ½ minutes) is long. Can you hold the breath for 90 seconds? I don't know if this is the actual reason for the 90-second rule, but it is a plausible explanation.

The second example is the duration of power interruption of 200 ms. Many ARINC documents mention the 200 ms rule. Why 200 ms? Why not 300 ms or 100 ms? A power supply that can maintain voltage for 300 ms after power interruption can be designed—as well as a power supply that can, during a power interruption, maintain voltage for 100 ms. So why was the duration of 200 ms chosen? It would be interesting to *Know-Why!* See if you can find the answer to the second example. If you are designing a new aircraft, system, or LRU, it is crucial to know *why* regulations and requirements are what they are. Even if you are redesigning the aircraft (737MAX or A320NEOS), it is crucial to *Know-Why* some things are designed the way they are designed. In the lifetime of 737 and A320, there will be upgrades where design engineers will scratch their heads and wonder *why*. Knowing why is sometimes very important, but the answer may be hidden in the heads of a select few. If they never take the time to write it down and share the knowledge, it will be lost forever.

Know number three is *Know-Where*. Know-Where means: Where can it be found? Know-Where is of paramount importance. When you forget the past, you are doomed to repeat it. This is also the point at which we must discuss how we have stored data. If you can't find *where*, you are doomed to redesign it, to reinvent something that may have been invented many years ago. The way we now store essential information enables us to search using keywords and a computer—which is a huge improvement. But we still have a limited number of people with vital knowledge who also know *where* the information is stored. When they are gone (due to retirement, winning the lottery, etc.), no one will Know-Where their documents are, which may be necessary to Know-Why something is done the way it is done.

I am certain that Know-How, Know-Why, and Know-Where can be preserved using the Internet and intranet. Unfortunately, due to the Coronavirus pandemic, many companies pushed people out to lower costs. Three years later (or sooner), those people and their knowledge will be missed. The culture of transferring Know-How, Know-Why, and Know-Where should be established in every industry to prevent the loss of institutional knowledge and the costs of re-creating it.

This is how I came to the end of my 40-year career in aviation. I carried the torch for 40 years: I was the Torchbearer. If somebody asked me, I would certainly say that the years between 1980 and the Coronavirus pandemic (2020) were the best years in aviation. I consciously enjoyed nearly every moment of that time. From my standpoint, I took the torch back in 1980 from engineers who were working on DC-8s and DC-9s. After that, I did my best for 40 years to enjoy the art of engineering and maintenance in one of the world's leading aircraft companies. I did a lot, learned a lot, and had a lot of fun. As a bonus, I have seen much of the world, and I have visited all the continents except Antarctica. Now, I am passing the torch on to a new generation of engineers. My shop engineers knew how I felt, and they organized a farewell meeting for me. I left my airline after 40 years and handed over the symbolic torch to my son, a chemical engineer. He is now working for the same company where I spent my career. I was really touched by that gesture from my dream team. It is deeply gratifying to have had such a career and to feel so appreciated. Let the youngsters carry the torch now. Let them be the Torchbearers.

FIGURE 25.2 Having a great time with ATEC6 series test computers.

Courtesy of Marijan Jozic.

Will I miss it? Absolutely not. I have seen it all, and for me, it would only be the same old same old. The world is limitless, and soon there will be a new beginning, even before I realize it.

About the Author

Marijan Jozic has worked in avionics engineering and maintenance for more than 40 years. He earned his Bachelor of Science in Aerospace Engineering in 1978 from the Higher Technical School at the University of Zagreb (Croatia) and began his career with KLM Royal Dutch Airlines in 1980. His first job was as Test Equipment Calibration and Maintenance Engineer in the Test Equipment Maintenance Laboratory. In 1986, he earned his second degree, a Bachelor of Science in Electronics and Telecommunications, from Higher Technical School at the University of Amsterdam (The Netherlands). In 1987, he joined KLM Engineering as an Avionics Systems Engineer. In 1998, he was promoted to Senior Engineer for Avionics Systems for 747 aircraft. During his career at KLM Engineering, he was Project Manager for several avionics fleet-wide projects, including SATCOM, RVSM, FM Immunity and 8.33 kHz channel spacing, and many, many more. In 2001, he was moved to the position of Program Manager for fleet-wide avionics projects, and a year later, he was promoted to Program Manager for all modification programs for the KLM fleet. In this position, he successfully completed many projects (glass cockpit on a 747 classic, TRAS lock, enlarged pitch, etc.). His greatest success was the project Avionics Master Plan, which was set up to install the TAWS/MMR/GPS/FMS system on the entire KLM fleet. For his work on this project, Marijan Jozic was awarded the Volare Award in Montreal at the AMC (Avionics Maintenance Conference) in 2004. The Volare is the highest award an avionics engineer can receive for "Significant Individual Achievement." The award is given by the AAI (American Avionics Institute). Marijan was greatly honored to receive the Volare, particularly because the recipient is chosen by his peer engineers.

Marijan was nominated twice (2004 and 2005) for the Aerospace Journalist of the Year award in the avionics category.

At KLM, Marijan and his avionics engineering group received the TQP (Total Quality Performance) award for the Avionics Master Plan Project.

In 2006, Marijan became a member of the steering committee of the AMC. At the AMC in Anchorage Alaska (2012), Marijan was elected chairman of the AMC, a position he held for 7 years. This was the longest period any individual served as chairman in the 70-year history of AMC.

In 2017, Marijan received the Roger S. Goldberg Award for endless passion for avionics maintenance. This award was especially meaningful because Roger S. Goldberg helped Marijan rediscover himself as an engineer-storyteller.

Marijan's final conference as AMC chairman was in Prague in 2019. At that conference, Marijan received the "Man of the Decade Award." This was the crowning achievement of his career in aviation maintenance.

Shortly after the COVID-19 pandemic began, the aviation industry collapsed, and Marijan accepted a voluntary retirement package from KLM. This doesn't mean that he stopped being active in aviation; his activities just shifted to SAE ITC (Industry Technology Consortia), which opened a window to a new aviation career.

Presently, Marijan serves as Director of European Operation at OctonX. OctonX is the official affiliate of SAE ITC, representing SAE ITC in Europe. Marijan's OctonX activities are divided among the aviation, automotive, and software industries.

Marijan lives in the Netherlands, and has been married to Connie since 1980 (Figure 1). They have two sons (Jan-Willem and Robert-Jan) and two grandchildren (Jo-Anne and Ivan).

FIGURE 1 Marijan and Connie Jozic.

Courtesy of Marijan Jozic.

CPSIA information can be obtained
at www.ICGtesting.com
Printed in the USA
BVHW010244100323
659961BV00029B/1297